Dictionary of
Marine Technology

Dictionary of Marine Technology

D. A. Taylor

MSc, BSc, CEng, FIMarE, FRINA
Senior Lecturer in Marine Technology, Hong Kong Polytechnic

Butterworths

London Boston Singapore Sydney Toronto Wellington

PART OF REED INTERNATIONAL P.L.C.   221084

First published 1989

© Butterworth & Co. (Publishers) Ltd, 1989

British Library Cataloguing in Publication Data

Taylor, D. A. (David Albert) *1956–*
    Dictionary of marine technology.
    1. Marine engineering
    I. Title
    623.87

ISBN 0-408-02195-0

Library of Congress Cataloging in Publication Data

Taylor, D. A., M.S.c.
    Dictionary of marine technology.

    1. Marine engineering—Dictionaries.   I. Title.
VM600.T38   1089   623.8′03′21   88-30419 •.

    ISBN 0-408-02195-0

Photoset by Butterworths Litho Preparation Department
Printed and bound in Great Britain by Anchor Press Ltd, Tiptree, Essex

Preface

Every industry has its specialist terms or particular use of well-known words. The marine industy is no exception. This volume attempts to explain terms related to marine and offshore engineering, naval architecture, shipbuilding, shipping and ship operation. Electricity, electronics, control and computing terms, where relevant to propulsion systems and ship operation, have been included. Terms printed in *bold italic* indicate entries that provide further information. Illustrations have been used in many instances to aid explanations.

This text is intended to replace the *Dictionary of Marine Engineering and Nautical Terms* written by the late G.O.Watson. While some of the present material has been drawn from this source the emphasis is now very much upon marine technology with only a few nautical terms included.

I quote from the Preface of G. O. Watson's dictionary:

'A technical dictionary should fulfil at least two functions. It should first be a source of information on words and expressions that are strange to us . . . Second, by suitable grouping and the use of cross references it should help the user to become familiar with groups of terms associated with a particular concept'.

I have endeavoured to produce such a dictionary and acknowledge the assistance received from this earlier volume.

The Appendix contains a list of marine related abbreviations and also a summary of the SI system of units with some conversion factors. Anyone with an interest in the technical or commercial aspects of ships and offshore floating structures should benefit by reading or referring to this book. After more than twenty years in the business I certainly learnt a lot researching and writing it!

D.A. Taylor

Acknowledgements

I would like to thank the many firms, organizations and individuals who have provided me with assistance and information used in the compiling of this dictionary.

The following firms provided information and illustrations, for which I thank them.

Asea Brown Boveri Ltd
Bank of Scotland
Ben Line Containers Ltd
Brown Brothers & Co., Ltd
Chadburn Bloctube Ltd
Frank Mohn A/S
Gusto Engineering c.v.
NEI International Combustion Ltd
Schottel-Werft
Swire Pacific Offshore
The Engineering Council
Young and Cunningham Ltd

A final thank-you goes to Miss Patience Ho Wai Mui, who typed my manuscript.

A

A-bracket A bearing support assembly for twin screw propeller shafts which project from the hull forward and on either side of the stern frame. See *spectacle frame*.

A-class divisions *Bulkheads* and *decks* which are constructed of steel or equivalent material and suitably stiffened. They must be constructed to prevent the passage of smoke and flame for a one-hour *standard fire test*. Finally they must be insulated such that the unexposed side will not rise more than 139°C or any point more than 180°C above the original temperature within times as follows: Class A-60, 60 minutes; A-30, 30 minutes; A-15, 15 minutes; A-0, 0 minutes. See *main vertical fire zones, B-class divisions, C-class divisions*.

A-frame A fabricated steel structural element of a slow speed, two-stroke, diesel engine. It stands on the *bedplate* above the main bearing. See Figure S.11.

A1 (1) A *class notation* awarded by *Lloyd's Register of Shipping*. (2) In first class condition.

abandoned well A *well* from which all equipment has been removed and the opening sealed.

abandonment The release of control over a vessel which is considered to be a *constructive total loss*.

abrasion (1) A mechanical action between two substances resulting in wear. (2) The scraping or scoring of gear teeth due to impurities in the lubricating oil and the sliding of teeth over one another.

absolute humidity The amount of water present in a given volume of air. It is usually expressed in grams per cubic centimetre. See *relative humidity*.

absolute temperature A measurement which is related to absolute zero, the theoretical minimum *temperature* for any substance. The unit is the kelvin (K) and conversion to the *Celsius scale* is by the addition of 273.15.

accelerometer An acceleration measuring instrument which can also be used to obtain measurements of displacement and velocity. It may also be used for vibration and shock measurements.

acceptance tests A series of tests performed on a material, a machine or a system, in the presence of the purchaser or a *Surveyor* to demonstrate suitable quality or performance.

accommodation The part of a ship's structure which is used for cabins, dining areas and other crew facilities. See Figure A.1.

accumulator (1) See *battery*. (2) A container for the storage of liquids and gases under pressure which acts as a reservoir.

accumulation test A boiler test to ensure that the *safety valves* can release steam fast enough to prevent the pressure rising by 10%. The *main steam stop valve* is closed during the test.

1

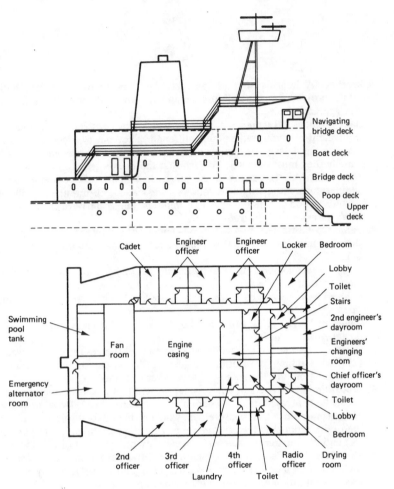

Figure A.1 Accommodation arrangement

accuracy The degree of closeness with which the indications of an instrument approach the true values of the quantities measured.

acetylene A colourless, poisonous gas which is used with oxygen for oxy-acetylene welding or cutting of metal. It is produced by the action of water on calcium carbide or high temperature cracking and solvent extraction from feedstocks such as naptha.

acidic attack   Metal damage which takes place in the gas path of a boiler due to acidic deposits, usually sulphur, which are absorbed by water to produce sulphuric acid.

acoustic (sonic) log   A device which is towed behind a vessel and directs sound waves to the sea bed. The time taken for reflection is used, with other data, to determine the rock structure of the sea bed.

Act of God   Any unforeseen act which could not have been avoided by suitable care and forward planning.

actual total loss   A form of *total loss* which indicates that the insured item has been destroyed, damaged or rendered irrecoverable such that it no longer exists.

actuator   A motor providing rotary or linear motion.

ad valorem   According to value. Freight rates, insurance quotations or other charges may be made on this basis.

added virtual mass   The mass of water which is considered to be set in motion by a ship when heaving, pitching, rolling or vibrating.

addendum   The distance from the pitch circle to the tip circle on a gear wheel.

additives   Chemicals which are added to fuel or lubricating oils to improve their physical or chemical characteristics.

address   A binary number that specifies a particular memory location in a *computer*.

adiabatic   The expansion or compression of a gas in which no heat flow occurs across the boundaries of the system.

admiralty coefficient   A value, determined from a simple formula, which is used to compare the performance of similar ships

$$\text{Admiralty Coefficient} = \frac{\Delta^{2/3} \, V^3}{P}$$

where: Δ = displacement in tonnes, V = speed in knots, P = shaft power in kW. Values range from 400 to 600, the higher the value the more economic the vessel.

admittance   The ratio of the *root-mean-square* values of *current* and *voltage* in a circuit. The reciprocal of *impedance*.

advance   The forward distance travelled by a ship, measured from the centre of gravity, parallel to its original course, from the moment when the *rudder* is moved hard over. See Figure T.9.

advance freight   A payment of *freight* charges in advance of the delivery of the cargo.

aerobic bacteria   Organisms which live and grow only in the presence of oxygen. They are used in biological *sewage treatment* plants.

aft   In a direction towards the *stern*.

aft peak   A *compartment* located aft of the aftermost watertight *bulkhead*. It is often a water storage tank.

after burning   Combustion which continues in diesel engines after valves or exhaust ports have opened or in the uptakes of boilers.

3

after perpendicular An imaginary line drawn perpendicular to the *waterline* of a ship where the after edge of the *rudder post* meets the summer *load line*; if no rudder post is fitted the centreline of the rudder pintles is used.

after shoulder The part of a ship's form where the *parallel middle body* and the *run* meet.

against all risks Cargo may be considered insured against all risks but this means only generally accepted risks and is therefore a limited cover.

aground Not floating freely. A part or all the vessel is supported by the ground.

air compressor A machine which provides cool compressed air which is initially stored in air receivers or 'bottles'.

air conditioning The control of *temperature* and *humidity* in a space together with the circulation, filtering and refreshing of the air.

air cooler A *heat exchanger* to cool and therefore increase the specific density of air prior to combustion in an engine. Finned tubes are circulated, usually with sea water, and have the air passing over their outer surfaces.

air dryer A unit used to remove moisture from *instrument air*. It may use a *desiccant* or a *refrigerant drier*.

air ejector A gas extracting device which is connected to *surface condensers*. The steam operated *ejector* draws out any gases released during the condensing process. The unit is circulated with feed water which is heated as it condenses the ejector steam. See Figure A.2.

air heater A *heat exchanger* which uses steam, water or electricity to increase the air temperature. Tubes are circulated with the heated medium and the air passes over their outer surfaces. Air heaters in boilers use exhaust gas and metal plates, or some other arrangement, to achieve heat transfer.

air lock A quantity of air in a pipeline which stops the flow of liquid.

air pipe A ventilation pipe fitted to liquid storage tanks. It enables filling or emptying to take place without significant pressure changes within the tank.

air receiver A vessel for storing compressed air.

air register See *register*.

alarm monitoring A computer controlled scanning process in which a measured value is obtained and then compared with set limits. If a condition occurs which is outside the set limits then audible and visual alarms will be given.

algorithm A defined sequence of operations that leads to the solution of a problem.

alignment gauge A measuring instrument which is used to check the alignment of an engine crankshaft, e.g. a *bridge gauge* or a micrometer clock gauge.

alkaline cell The basic unit of an alkaline battery. Marine alkaline cells are usually of the nickel cadmium type. The positive plate is nickel hydroxide

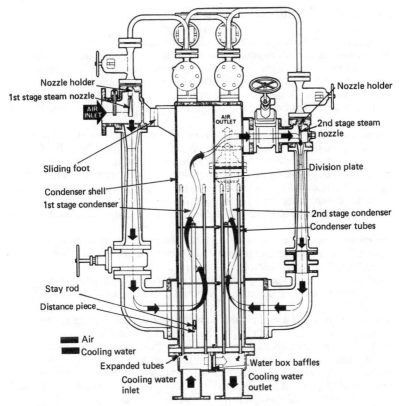

Figure A.2 Air ejector

and the negative plate is cadmium and iron. The electrolyte is potassium hydroxide.

alkalinity A measure of the alkaline nature of a solution. The *pH* value of alkaline solutions ranges from 7 to 14, the higher number indicating strongly alkaline.

alkyd resin A hard, tough, long lasting material which adheres well to most surfaces. It is used in many types of priming and undercoating *paints*.

alphanumeric A set of characters which includes letters, numbers and some punctuation marks.

altar The ledge in a *dry dock* where ship side *dog shores* are positioned.

alternating current An electric *current* which alternately increases and decreases in a periodic manner. The frequency of the alternating cycle is independent of the circuit constants.

alternator An alternating current *generator*.

aluminium A light metal which has a good resistance to atmospheric corrosion. It is used as the base metal for light alloys.

aluminium alloys *Aluminium* alloyed with materials such as copper, manganese, silicon or magnesium, to improve its strength for use in structures such as the accommodation in passenger ships.

aluminium brass *Brass* containing up to 6% *aluminium* in order to improve resistance to corrosion. It is used in tube plates and tubes of condensers and heat exchangers.

aluminium bronze A copper aluminium alloy containing 4–11% *aluminium* and other elements for particular properties. It is used for valve fittings and other applications where corrosion resistance is important.

always afloat A *charter party* requirement that at all stages of the tide a ship must remain afloat. The *charterer* must therefore ensure an appropriate water depth exists at all *berths*.

American Bureau of Shipping The *Classification Society* of the Unites States of America.

American Petroleum Institute A trade association which has produced many specifications and procedures which have become accepted almost as standards by the petroleum industry.

amidships The middle region of a ship between *stem* and *stern*. When refering to steering the rudder is in mid-position, (*midships*). See Figure P.7.

amine A chemical used in boiler *water treatment*. Neutralizing amines such as cyclohexylamine and morpholine are used to neutralize carbon dioxide. Filming amines such as octadecylamine are used to prevent corrosion by producing a water-repellant film on metal surfaces.

ammeter An indicating instrument used to measure electrical *current* flow.

ammonia A pungent smelling noxious gas which is very soluble in water and produces an alkaline liquid. It has been used as a refrigerant.

ampere A base unit for the measurement of electric *current*.

amplification The ratio of output to input magnitude in a device which is designed to produce an increased value output.

amplifier A device in which an input is used to control a local source of power so as to produce an output which is greater than, and bears a definite relationship to, the input.

amplitude The maximum displacement of a varying quantity, measured from some datum.

analogue computer A *computer* which uses data in the form of continuously variable physical quantities. The data, representing for example a pressure or a temperature, is transduced into an electrical quantity which is an analogue of the data.

anchor A device for mooring a floating vessel; one end digs into the sea bed and the other is attached to a cable by a ring.

anemometer An instrument used to measure wind speed. It may also indicate direction.

6

aneroid barometer   An *atmospheric pressure* measuring instrument which uses the movement of an evacuated cylinder or bellows as the transducer or sensor. A series of levers or linkages moves a pointer over a scale to give a measurement.

angle of incidence   The attitude of a blade or aerofoil section to the fluid flow.

angle of loll   The angle to which a ship, which is initially unstable and wall-sided in the region of the waterline, will *heel* or *list*. At this angle it will be stable.

annealing   A heat treatment involving heating to a particular temperature and then cooling at a suitable rate. *Ductility* is improved and the material can be more easily worked when cold.

anode   (1) The *electrode* through which a *direct current* can enter a liquid or gas. (2) The main electrode for the collection of electrons (*electronics*).

anti-fouling paint   A *paint* which functions to inhibit the adhesion of marine life to underwater surfaces. The paint leaches out a poison into the water.

anticline   A fold in layers of rocks where the strata slope down at either side. Oil and gas are often found in such structures. See Figure A.3.

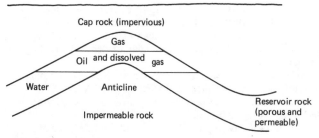

Figure A.3 An anticline

aperture   The open area formed by the *stern frame* and *rudder*, in which the propeller operates.

API gravity   Another term for *degree API*.

apparent good order and condition   A shipowner accepts cargo in this stated condition because he has no right to inspect the contents.

appraisal well   A hole drilled near to a *discovery well* in order to determine more information or the extent of the *reservoir*.

arbitration   A process whereby matters in dispute are submitted to an agreed independent person or persons for judgement without involving the Law Courts.

arc welding   The fusing of two metals using the heat produced by an electric arc formed between them. Various different welding processes use the basic principle of an electric arc. See *metal inert gas welding*.

7

argon An *inert gas* which is used, for example, to shield the molten metal during argon arc welding.

arithmetic and logic unit A part of the central processing unit of a *computer* where arithmetic and logic operations are performed.

armature The part of an electrical machine which carries the winding which is connected to an external supply and in which the main e.m.f. is induced. It is usually the rotating part of a d.c. machine but on an a.c. machine it would be static and probably called the *stator*.

armouring Mechanical protection provided over the electrical insulation of *cables*. A basket-woven wire braid of either tinned phosphor bronze or galvanized steel may be used.

arrest The detention of a ship until the purpose of the arrest has been achieved.

articulation A connection or joint between two components. When referring to levers or gears the loads are usually equalized or shared by appropriate proportioning of the leverages with respect to the fulcrum.

as fast as can When a ship is loading cargo the charterer is obliged to supply the cargo as fast as the ship can load it, bearing in mind the handling capacity of the port.

asbestos An *incombustible* material which, in a fabric or wool-type form may be used for thermal insulation, i.e. lagging of equipment and piping. Some types may release dust which is a health hazard.

aspect ratio The ratio of the span to the mean chord line of an aerofoil section. For a ship's rudder this is the mean span or depth divided by the mean chord or width.

assemble The preparation of a *machine-language* program from a *program* written in a symbolic language such as *BASIC* or *FORTRAN*.

assembly A large, usually three-dimensional, structure of plating and sections used in the construction of a ship.

astern (1) The position of any ship or object behind the reference ship. (2) The direction of rotation of the main propulsion engine which produces a stern-first motion through the water.

astern turbine A short length of nozzles and blades on a turbine rotor which will provide about 30% of full power to drive the vessel astern.

athwartships In a direction across the *breadth* of a ship.

atmospheric drain tank A tank in a steam plant feed system which collects condensed steam from various sources, e.g. drains. The tank is open to the atmosphere.

atmospheric pressure The pressure exerted upon the surface of the Earth. The standard value is 1.01325 bar.

atomization The breaking up of a material into very small parts, in particular a liquid into small droplets or a fine spray.

atomizer A *nozzle* arrangement through which *fuel oil* is forced under pressure in order to leave as a fine spray i.e. atomized.

attemperator A *heat exchanger* which cools or desuperheats in order to control the superheat temperature of steam leaving a boiler.

attenuation A reduction in current, voltage, or power due to transmission. The opposite of *amplification*.

audio frequency The *frequency* range of pressure waves which produce sound, i.e. 25–20 kHz.

augment of resistance Additional *resistance* or form drag which must be considered when assessing the *frictional resistance* of a ship's hull by comparison with a flat plane of equivalent area.

austenitic steel *Steel* which has been alloyed with sufficient amounts of nickel, chromium and nickel, or manganese to retain austenite at atmospheric temperature.

auto-kleen filter A self-cleaning *filter* consisting of a wire spirally wrapped around a perforated drum. Small metal leaves are fitted between adjacent wires in a vertical column. The wire wound drum can be rotated externally by hand or an electric motor and any filtered matter will be collected by the metal leaves and fall to the sump of the filter.

auto-transformer starting A method of reduced voltage starting for an *induction motor* to limit the starting current. The *transformer* used has only one winding for both input and output.

automatic controller An element in an automatic *control system* which receives a signal representing the *controlled condition*. This is then compared with a signal representing the *command signal*. The output signal then operates to reduce the *deviation*.

automatic valve The *valve* used to admit starting air into the cylinder of a diesel engine. It is open only when the starting handle is in the 'air' position and is automatically closed in the 'stop' and 'fuel' positions.

automatic voltage regulator Electrical equipment which senses and controls an a.c. *generator's* output voltage to within plus or minus one or two per cent.

auxiliary blower An electric motor driven fan which is used to supplement the *scavenging air* to a diesel engine during low speed operation.

auxiliary boiler A *boiler* which supplies steam for essential auxiliary purposes but not for main propulsion. It is often of the *fire tube* type. See *package boiler*.

auxiliary steam Low pressure *saturated steam* for various services such as pump operation, fuel heating and domestic water heating.

average A loss of either a general nature, i.e. *general average,* or a particular nature, i.e. *particular average.*

average adjuster An assessor who will determine losses and contributions required under a *general average* act. An authority on marine insurance law and loss adjustments.

axial flow pump A *pump* which is effectively a *propeller* in a tube. A large volume flow of liquid at low pressure is produced and used usually for condenser circulating on steam turbine plant. See Figure A.4.

azimuth thruster A *thruster* which can rotate through 360 degrees. It may be retractable or fixed in position.

9

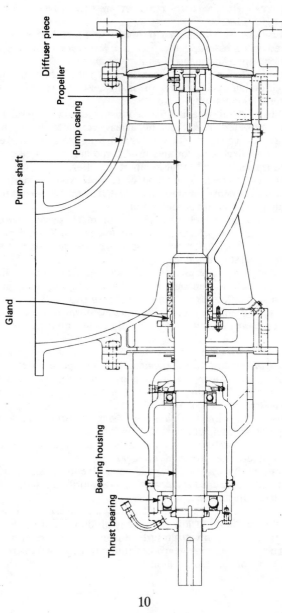

Diffuser piece

Propeller

Pump casing

Pump shaft

Gland

Bearing housing

Thrust bearing

Figure A.4 Axial flow pump

10

B

B-class division A *bulkhead* constructed to prevent the passage of flame for a half-hour *standard fire test*. It must be insulated so that the unexposed side will not rise more than 139°C or any point 225°C above the original temperature within times as follows: Class B-15, 15 minutes and B-0, 0 minutes. See *A-class divisions, C-class divisions*.

Babbit's metal A metal with low friction properties, used to line *bearings*. It is an alloy of tin, copper, antimony and lead.

back pressure The *pressure* existing on the exhaust side of a system, e.g. the pressure opposing the motion of an engine piston on its exhaust stroke.

backfreight The *freight* charged for the return of goods.

backlash The motion lost between two elements of a mechanism due either to working clearances or wear.

baffle plate A plate used to direct fluid flow, e.g. the hot gases in a boiler furnace or the oil in a sump tank.

balance (1) Electrical balance refers to *bridge* measurements and is achieved when the *impedance* of each arm has been adjusted so that no *current* flows through the *galvanometer*. (2) Dynamic balance is achieved in a moving object or assembly when the motion is in equilibrium and without vibration from any reciprocating motion or centrifugal force. (3) Static balance is achieved when an object is at rest and in equilibrium about an axis or point.

balanced rudder A *rudder* design in which more than 25% of the area is forward of the turning axis. There is therefore no torque on the *rudder stock* at certain angles. See Figure B.1.

balancing A means of reducing vibrations in a reciprocating engine by balancing out forces and couples or reducing their effects to acceptable values.

ballast (1) A heavy substance, usually sea water, taken on board a ship to increase the *draught,* change the *trim* or improve the *stability*. (2) The condition of a ship when no *cargo* is being carried. Some tanks are filled with sea water to make the ship seaworthy.

bandwidth The *frequency* range within which certain harmonic response characteristics, such as *gain* and *phase angle,* are within specified limits.

bar (1) A bank of silt across a river mouth. (2) A unit of *pressure* equal to 100 000 *pascals*.

bareboat charter A *charter party* where all expenses during the period of hire are paid by the *charterer*.

barge carrier A vessel designed to carry standard barges which have been loaded with cargo. The barges, once unloaded, are towed away by tugs and return cargo barges are loaded. Two particular types are *LASH* (Lighter Aboard Ship) and *SEABEE*.

11

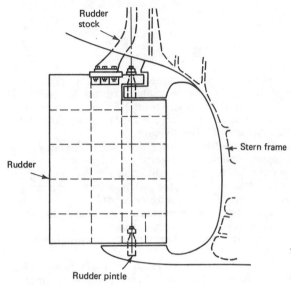

Figure B.1 Balanced rudder

barometer An instrument used to measure *atmospheric pressure*.

barratry Any fraudulent act by the master or crew of a vessel acting against the interests of the shipowner or charterer, e.g. scuttling or smuggling.

barrel A volumetric measurement of *petroleum* and its products. It is equal to 42 US gallons and approximates to 35 Imperial gallons or 159 litres. It may be written as b, bbl or brl.

base The number from which a numeration or logarithm system begins, e.g. 2 in a *binary number system*.

base line A reference line extending *fore and aft* along the upper surface of the flat plate keel at the centre line.

BASIC Beginner's All-purpose Symbolic Instruction Code. A symbolic high-level computer programming language.

battery A number of secondary *cells* connected in a circuit to provide electrical power. The term *accumulator* is also used.

baud A unit of transmission speed for computer telecommunications of one bit per second. It is often used in computing to indicate the rate at which a printer will accept information. To convert a baud rate into an approximate number of characters per second, divide by 10.

beam A rolled steel section supporting the *deck* and positioned athwart-ships.

beam knee A metal plate joining the *deck beam* and the *frame* at the ship's side. See Figure F.4.

12

bearing A common item in any mechanical system where two parts move relative to one another. It enables the transfer of forces with a minimum of frictional losses.

bedplate A structure which forms the base of a machine upon which the bearings and frame are mounted. See Figure S.11.

bellmouth A suction pipe arrangement used in a tank. The inlet area is increased to about one-and-a-half times the pipe area.

bellows A thin metal cylinder the walls of which are corrugated to permit reasonable deflection under the action of an applied pressure. It may be used in measuring instruments or pneumatic control equipment.

bellows expansion joint An expansion piece which is fitted into a pipeline to allow expansion and contraction during temperature changes.

bend test A test to measure the *ductility* of a metal sample by folding it over a specified radius. No cracking or other defects should be found.

bends A colloquial term for *decompression sickness* which may occur in divers who surface too quickly after working in deep water.

berth (1) The place at a *quay* or *pier* where a ship is to be moored. (2) The bunk or cabin provided for a passenger or member of the crew.

big end The larger end of a *connecting rod*, i.e. where it joins the *crank pin* of the *crankshaft*.

bilge (1) The curved portion of plating between the bottom shell and side plating. (2) A drainage space in a *hold* or a region of the *machinery space*.

bilge block A large wooden block which is used to support the bilge region of a ship when in *dry dock*.

bilge injection valve See *direct bilge suction*.

bilge keel A projecting flat plate positioned at right-angles to the radiused plating at the *bilge*. It extends for about one-half of the length of the ship. See Figure B.2.

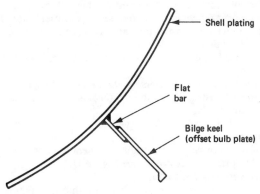

Shell plating

Flat bar

Bilge keel (offset bulb plate)

Figure B.2 Bilge keel

13

bilge pipe   The piping used for draining the bilge regions of a ship. The suction end is fitted with a *strainer* or *mud box.*

bilge pump   The *pump* which is used for extracting water from the bilges and pumping to the *oily water separator* and then overboard.

bilge radius   The radius of the plating joining the side shell to the bottom shell. It is measured at *midships.*

Bill of Health   A certificate issued by the medical officer of the port stating the condition of health of the ship or the port.

Bill of Lading   A receipt issued by the master or his agent for goods being shipped. It is a document of title and details the terms and conditions of carriage.

Bill of Sale   The document used to transfer the ownership of a British ship which must be registered.

bimetallic joint   Two different metals which are joined together in a structure. They must be electrically insulated one from another to avoid *galvanic corrosion,* e.g. a steel hull with aluminium superstructure.

binary number system   A number system using only 0 and 1 and having a base of 2.

bipod mast   A tied, or structurally connected, pair of *samson posts.*

bit   (1) The smallest unit of data in binary notation, it may be 0 or 1. (2) The cutting tool which is used to drill *bore holes.* See Figure D.2.

bitter end   The end of the anchor cable which is secured in the *chain locker* by the *clench pin.*

bitts   The vertical tubular posts, two of which are fastened to the rectangular base of a *bollard.* See Figure B.3.

black oil   A very heavy or dark *crude oil,* e.g. *fuel oil.*

blade area   The area enclosed by the developed blade outline of a *propeller.*

block coefficient   The ratio of the *volume of displacement* at a particular *waterline* to the volume of a rectangular section of the same length, *breadth* and *draught* as the ship.

block diagram   A number of blocks interconnected by 'ines. A block is a rectangle containing a *transfer function.* Lines are used to represent inputs to and outputs from the block and arrows indicate the direction of flow.

blowdown valve   A *valve* which enables water to be blown down or emptied from a *boiler.* It may be used when partially or completely emptying the boiler.

blower   A fan used to supply air which will *pressure charge* a diesel engine. It may be driven by an exhaust gas turbine (*turboblower*), electricity (*auxiliary blower*) or mechanically.

blowout   A sudden uncontrolled escape of oil or gas at high pressure from a *well* during the drilling stage.

blowout preventer   A collection of *valves* and other safety equipment which is fitted on a *well-head* to prevent a *blowout.* See Figure D.2.

Board of Trade   See *Department of Transport.*

body plan   A drawing showing half transverse sections of a ship. See Figure L.3.

14

body sections   The ship sections obtained by planes perpendicular to the centreline. They are drawn on a *body plan* and form part of the *lines plan*. See Figure L.3.

boil-off   The evaporation of *liquefied petroleum* or *natural gas* due to heat transfer into the cargo tanks.

boiler   A pressure vessel in which water is heated to produce *steam*. The two main types are watertube and firetube and various designs of each are in use.

boiler mountings   The various fittings which are attached to a *boiler* to ensure its safe operation, e.g. *safety valves*.

boiler water treatment   See *water treatment*.

bollard   A rectangular base upon which are welded two vertical posts or *bitts*. The bollard is securely welded to the deck of a ship and used to secure the mooring lines. See Figure B.3.

bonding   The electrical connection between the armouring or sheath of adjacent lengths of cable, or across a joint, to ensure continuity.

Bonjean curves   Lines or curves of immersed cross-sectional area plotted against *draught* which are often drawn on the profile of a ship. They enable the determination of underwater volume for *waterlines* which are not parallel to the base.

boom   (1) A long, seamless steel tube, hinged at its lower end and supported by tackle at its upper end. It is used for cargo handling and may be called a *derrick* boom. (2) A series of floating obstructions secured together to restrict access, e.g. around the after end of a twin screw ship, or to contain floating matter, e.g. oil.

boot topping   The area of a ship's side plating in the region of the *load lines*. It is the hull area which is most susceptible to corrosion.

borehole   A hole drilled in search of oil or gas.

borescope   An optical aid, also called an *endoscope,* to enable the examination of internal metal parts or surfaces via small access holes, e.g. *liners* and valve seats by inserting the device through the injector hole.

boss   The central portion of a *propeller* to which the blades are attached and through which the *tailshaft* passes.

bossing   A curved arrangement of a ship's *bottom shell plating* which supports the *propeller shaft* of a twin-screw ship.

bottom dead centre   The position of the *piston* in a *cylinder* when it is at the end or bottom of its *stroke*.

bottom end bearing   The *bearing* at the end of a *connecting rod* where it joins the *crankpin* of the *crankshaft*. It is also known as the *big end* bearing. See Figure T.7.

bottomry   A loan obtained against the ship's hull, equipment or cargo which is repaid, with interest, upon its safe return. The loan is forfeit if the ship sinks.

bottom shell plating   The plating of a ship between the *keel* and the *bilge radius*.

bottoms   The heavy oil residue left at the bottom of a still after *distillation*.

15

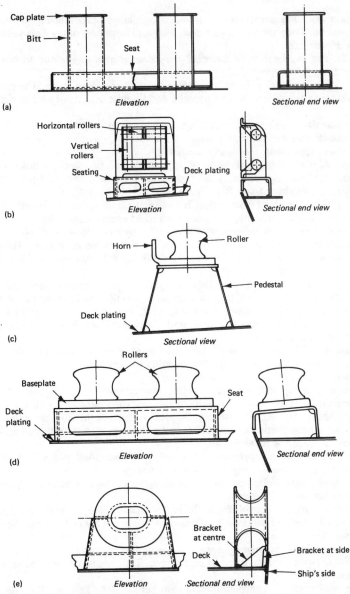

Figure B.3 Bollards and fairleads: (a) fabricated bollard, (b) multi-angle fairlead, (c) pedestal fairlead, (d) two-roller fairlead, (e) panama fairlead

boundary layer   A narrow layer of moving water adjacent to the hull of a ship as it moves through the water.

Bourdon tube   An elliptical section tube which is formed into a C-shape and sealed at one end. It is used as a measuring device for pressure and some temperature gauges.

bow   The forward end of a ship.

bow stopper   Another term for *cable stopper*.

bow thruster   A *thruster* fitted in an *athwartships* tunnel near to the *bow*.

bow visor   A hinged opening bow door where a part of the bow structure lifts to provide *roll-on roll-off* access to vehicle decks.

bracket   A flat metal plate, usually triangular, which is used to rigidly connect two or more structural parts, e.g. a *frame* to a *deck beam*.

branched system   A propulsion arrangement where there is a twin input or multiple output gearbox.

brass   An alloy of *copper* and *zinc* usually with a major proportion of copper. Various alloying metals may be added to improve properties, e.g. Admiralty brass contains a small amount of tin.

brazing   A means of joining metal items using a molten non-ferrous filler, which is applied to the joint. It forms a thin layer of metal which is fused to both surfaces.

breadth moulded   The breadth of a ship, measured to the inside edges of the *shell plating*. See Figure P.7.

break   The point at which a side shell plating section drops to the deck below, e.g. the *poop* or the *forecastle*.

break bulk cargo   General cargo which is packed in a variety of different containers or may be loose, e.g. boxes, drums, bales.

breakdown maintenance   A policy of plant repair only after it has failed. See *planned maintenance*.

breaking capacity   The ability of a *circuit-breaker* or other similar device to break or open an electric circuit under certain specified conditions.

breakwater   A vertical metal plate structure fitted across the *forecastle* or the *weather deck* to direct water which comes over the bows, off the deck and down the ship's side.

breast hook   A horizontal flat plate which stiffens the *stem* structure. The term breast plate is also used. See Figure B.4.

breathing apparatus   Equipment which enables a person to receive a supply of oxygen in an environment where little or no air exists, e.g. a smoke filled compartment. Two basic types are in use: the smoke helmet and the self-contained unit using air cylinders.

bridge   (1) The uppermost superstructure deck which extends to the ship's side, either as a fully enclosed structure or with open *bridge wings*. The ship is manoeuvred from this position which has a clear view forward. See Figure A.1. (2) The part of a valve body through which the spindle passes. It may also be called the *yoke*. (3) A method of connecting items in an electric circuit. See *Wheatstone bridge*.

bridge control   A control system which enables the remote operation of the

17

main machinery from the *bridge*. It must contain circuits to provide the correct logic, sequence and timing of operational events. There must also be protective devices and safety interlocks built into the system.

bridge gauge   A gauge which enables the measurement of journal bearing weardown with the crankshaft in place.

bridge rectifier   A full-wave rectifier which uses four diodes arranged as a *bridge*. The a.c. supply is applied across one diagonal and the d.c. supply is drawn from the other.

bridge wings   An open walkway from the *bridge* to either side of the ship.

brine   (1) A *refrigerant* produced by dissolving *calcium chloride* in water. (2) An excessively salty water, e.g. the liquid discharged from a boiling type *evaporator*.

brine trap   (1) A cylindrical chamber with appropriate fittings, for collecting samples of distiller or refrigerant *brine* for testing. (2) An S-bend in a hold *scupper* which contains *brine* and acts as an insulator.

British Marine Equipment Council   A trade association of British manufacturers of marine equipment.

British Maritime Technology   A national research organization with foundations in traditional fields of marine research and development. Its purpose is to provide high technology research, development, test services, products and consultancy.

British Ship Research Association   A British research organization, based in Wallsend, which is now part of *British Maritime Technology*.

British Standards Institution   A national organization which prepares and issues standard specifications.

British thermal unit   An Imperial unit of energy measurement. It is the heat required to raise the temperature of one pound of water from 60°F to 61°F. It is equal to 1055.79 joules. The term is still used in the oil industry.

brittle fracture   The fracture of metal with little or no plastic deformation. It occurs suddenly and at low stress levels and was a problem in early welded ships, particularly at low temperatures.

broach   To break open and steal, e.g. entering a container carried as cargo.

broken stowage   Space in a *hold* which cannot be filled because of the nature or type of the cargo.

bronze   An alloy of *copper* and *tin* with superior corrosion and wear resistance to *brass*. It is also alloyed with other metals to produce *aluminium bronze, manganese bronze* and *gunmetal* (adding zinc).

brush   A block of carbon which acts as a rubbing electrical contact on a *commutator* or slip *ring*.

bubble memory   A memory device in a *computer* which stores data even when the power supply is disconnected.

bucket valve   The suction and discharge non-return valves fitted on some reciprocating displacement pumps.

buckler   A sliding plate which covers the *hawse pipe* opening when the ship is at sea.

buckling   The bending or failure of a structural item when loaded by an excessive axial compressive stress.

18

buffer A temporary storage area in a *computer*.

bulb angle A rolled *steel section* which has one edge thickened with a bulb that is usually offset to one side. It is used as a stiffener for plating.

bulbous bow A cylindrical or bulb shaped underwater *bow* which is designed to reduce *wavemaking resistance* and any *pitching* motion of the ship. See Figure B.4.

bulk carrier A single-deck vessel which has been designed and constructed for the carriage of loose solid cargoes in bulk. The cargo carrying section of the ship is divided into holds, the arrangements of which vary according to the cargo to be carried. See *ore/bulk/oil carrier, ore carrier, ore/oil carrier*.

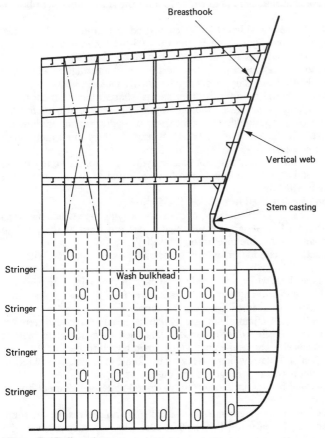

Figure B.4 Bulbous bow

19

bulkhead A vertical division arranged in a ship's structure to produce *compartments*. They may be transverse or longitudinal. Three basic types exist, *watertight,* non-watertight and *oiltight.* An oiltight bulkhead is watertight in construction but more rigorously tested than a watertight bulkhead.

bulwark A barrier fitted at the deck edge to protect passengers and crew and avoid the loss of items overboard should the ship roll excessively. They are considered solid or open. The solid type is of plate construction and the open type consists of railings.

bumpless transfer A smooth changeover between manual and automatic control or vice versa. It is usually achieved by having the manual and automatic signals follow one another during system operation.

bunker A *compartment* or *tank* in which fuel for the ship's engines or boilers is stored.

bunker fuel A heavy residual fuel which is used in marine diesel engines. The *kinematic viscosity* will be in the range 30–450 centistokes (cSt) at 50°C, 200–4500 seconds Redwood No. 1 at 100°F (37.8°C).

bunker receipt A document providing certain basic information regarding the fuel delivered to a ship, e.g. specific gravity and *viscosity*.

buoy A floating metal cylinder or other special shape which is anchored in position. It may serve as a mooring for a ship or a platform for the installation of navigational, meteorological or hydrographic equipment. Special types of tanker loading buoys also exist which are called *single point buoy moorings*.

buoyancy The vertical components of the hydrostatic forces acting on an immersed body. If the weight of the body is equal to, or less than, the buoyancy it will float.

burner The *atomizer* used in a *boiler*. See Figure R.3.

bury barge A barge with special equipment for burying pipe which has been laid on the sea bed. High pressure water jets are fired at the sand to fluidize it and enable the pipe to bury itself.

bus An electrical conductor which transmits power or data within a *computer*.

bus bar Copper bars which are fitted at the back of the main *switchboard* as part of the *distribution* system. *A.c. generators* feed to the bus bars and *circuit breakers* are used to draw off the supply.

butane A colourless, flammable gas which is a natural constituent of *petroleum*. It is easily liquefied and widely used as a domestic fuel. It improves volatility and anti-knock value when added to petrol.

butt strap A metal plate which covers a butt joint. It is usually welded to the adjoining plates to strengthen the butt joint.

butt weld A weld between the edges of two metal plates which meet but do not overlap.

butterfly valve A rotary stem *valve* with a centrally hinged disc of the same dimensions as the pipeline in which it fits. The valve opens into the pipeline and therefore takes up little space, permits large flow rates and

20

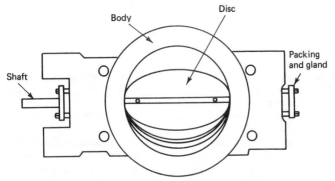

Figure B.5 Butterfly valve

gives minimum pressure drop. See Figure B.5.

Butterworth system A cargo oil tank cleaning system for a *tanker*. Hot sea water jets rotate within the tank to remove *crude oil* deposits. The resulting oily water mixture is pumped to a *slop tank* for separation.

butts The vertical welds in a ship's plating.

buttocks The after body sections of the *sheer profile* of a ship. These sections are the result of vertical planes parallel to the centreline intersecting the outer hull surface.

butyl rubber A synthetic *rubber* which is used for insulating electric cables. It has good heat, ozone and moisture resistance but is damaged by oil. It has largely been replaced by *ethylene propylene rubber*.

by-pass Any arrangement to control and divert a fluid from its main flow path.

by the head The longitudinal *trim* of a vessel when the draught forward is greater than the draught aft.

byte A group of adjacent *bits* which form a storage unit in a computer memory.

C

C-class division A bulkhead or deck which is constructed of non-combustible material, but meets no other requirements. See *A-class division, B-class division.*

cabin A compartment in the accommodation provided for the use of a passenger or member of the crew.

cable (1) A linear distance equal to one-tenth of a *nautical mile.* (2) The lengths of chain connecting the anchor to the ship. (3) A length of electrical conductor covered with insulation. Larger cables may be sheathed with lead or armoured with metal wire.

cable lifter A barrel with specially shaped 'snugs' into which the links of the anchor cable fit as they move round and drop into the chain locker. It is driven by the *windlass.*

cable ship A ship designed and equipped for laying and repairing subsea cables.

cable stopper A device used to hold the anchor cable in place while the ship is at anchor or the anchor is fully housed. See Figure C.1.

caisson A component part of an offshore structure which is watertight and made of precast concrete or steel, e.g. the base of a gravity structure.

calcium chloride A chemical which when mixed with fresh water is called brine and is used as a *secondary refrigerant.* It may be used as a drying agent in primary *refrigerant driers.*

calibration The process by which the readings of an instrument are compared with some standard or known value.

calorific value The quantity of heat released when one unit mass of a fuel is completely burned. The *higher calorific value* is the heat energy resulting from combustion. The *lower calorific value* is a measure of the heat energy available and does not include the heat energy contained in steam produced during combustion which passes away as exhaust.

calorifier A fresh water heater employing steam or electricity as the heating medium.

cam A shaped projection on a rotating shaft which imparts a motion, usually linear, to a follower in contact with it. See Figure T.7.

camber The curvature of the deck in a transverse direction. It is measured between the deck height at the centreline and at the ship's side. See Figure P.7.

camshaft A shaft fitted with one or more cams and driven by some mechanism from the crank shaft. See Figure T.7.

cant beam A deck beam joining a cant frame as part of the structure of a *cruiser stern.* It is considered cant in that it is set at an angle to the centreline of the ship.

cap rock A layer of rock which is impervious to oil or gas. It overlies a rock

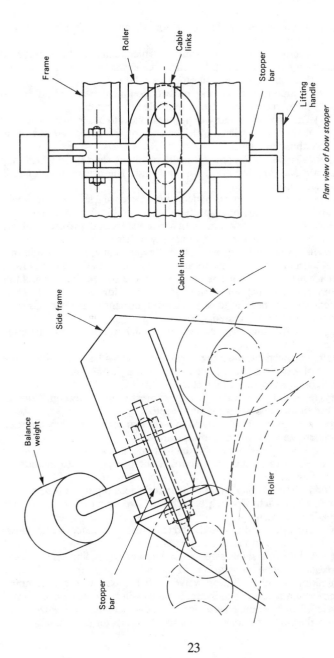

Frame

Roller

Cable
links

Stopper
bar

Lifting
handle

Plan view of bow stopper

Balance
weight

Side frame

Cable links

Stopper
bar

Roller

Figure C.1 Cable stopper

23

formation where a reservoir is to be found, and prevents the upward migration of hydrocarbons. See Figure A.3.

capacitance The property of a *capacitor* which enables the storage of electrical energy when a potential difference exists across its conductors, i.e. $C = Q/V$. The unit of measurement is a farad.

capacitor A component consisting of two conductors separated by an insulator which creates *capacitance* in an electric circuit.

capacity (1) The ratio of quantity to potential used to describe a controlled system, e.g. thermal capacity. (2) The sum of the *swept volumes* of an internal combustion engine.

capstan A device consisting of a vertically revolving shaft upon which may be fitted a *cable lifter* and also a warping end. The driving machinery is located below deck.

Captain's Protest A declaration made by the Master of a ship, giving details of any accident, damage or suspected damage to his ship or cargo.

carbon arc welding The use of one or two carbon electrodes to create an arc which will melt the filler metal and usually a flux.

carbon dioxide A colourless gas which is heavier than air. It is stored in a liquefied form in fire extinguishers and is sometimes used as a refrigerant.

carbon tetrachloride A colourless liquid which is a solvent for oils and fats and does not conduct electricity. It is used for cleaning electrical equipment and has been used as a fire extinguishant. It is toxic and must only be used in well ventilated spaces.

carburizing A hardening process for *steel* which produces an increased carbon content surface layer.

cardan shaft A mechanical arrangement which provides flexibility in the alignment of a driving and a driven shaft, i.e. a *flexible coupling*.

cargo The goods transported by a ship.

cargo carrying capacity A value usually expressed as deadweight tons or tonnes but other units are sometimes used.

cargo clusters A group of lights fitted in a circular reflector. They are used to illuminate the deck when working cargo.

cargo heating coils Pipes through which steam is passed to heat liquid cargoes in order to reduce their viscosity for pumping or to maintain a particular temperature.

cargo manifold The terminal point of a tanker's deck piping. It consists of a number of pipes each of which branch into two or more open ends for cargo loading or discharge.

cargo plan A diagram of ship's holds and cargo carrying spaces which shows the location of the various items of cargo.

cargo port A door in a ship's side for loading and unloading cargo.

cargo segregation The separation of liquid cargoes to avoid cross contamination. Double shut-off valves will be used where a common pumping system is used or individual pumps will be fitted in each tank.

Carnot cycle The operating *cycle* for an ideal heat engine which gives maximum thermal efficiency. It consists of an isothermal expansion, an

24

adiabatic expansion, an adiabatiac compression and an isothermal compression to return to the initial state. All the operations are reversible.

carrier wave A waveform, used in a signal transmission system, which may have its frequency, amplitude or phase varied by *modulation*.

carryover The passage of water from a boiler into the steam range and thus into machinery.

cascade control system A control system wherein one controller (the master) provides the command signal to one or more other controllers (slaves).

casing (1) A plated structure which separates the *machinery space* from the *accommodation*. See Figure A.1. (2) Steel pipes which line a well in order to prevent the borehole collapsing. They also prevent leakage into or out of the borehole. See Figure D.2.

cast iron Iron in which the carbon content may vary from 1.8 to 4.5%. White cast iron is hard and brittle. Grey cast iron is softer, readily machinable and less brittle. The colour refers to the appearance of the fractured surface.

casting (1) The pouring of molten metal into a mould of the desired shape. (2) Any item produced as in (1) .

catalytic cracking A secondary refining process which converts the larger molecules of heavy oils into smaller molecules of high grade gasolines by reaction with a catalyst.

catalytic fines Small particles, or fines, of the catalyst used in catalytic cracking. They are very hard, abrasive alumina and silica particles which are present in residual fuels. They must be removed by suitable treatment to avoid damage to pumps and injectors.

catamaran A vessel with twin hulls and a deck structure between them.

catenary The curve produced by a uniform, flexible, wire or chain when suspended by its ends. Anchor chains from a buoy or a towing wire between vessels will assume this shape. It provides resilience to any sudden stresses.

catenary anchor leg mooring A type of *single point buoy mooring* system where a flat cylindrical buoy is secured by a number of *catenary* anchor chains. It acts as a tanker loading platform receiving oil via sub-sea hoses from the terminal.

cathode An *electrode* which forms the negative terminal of an electrolytic cell. See *electrochemical corrosion*.

cathode ray tube A display device forming part of an *oscilloscope* or a television screen.

cathodic protection The prevention of *electrochemical corrosion* by providing an electrolytic cell in which the protected metal is the cathode. The anode may be formed by a metal which is sacrificed or an impressed d.c. current.

catwalk A narrow unfenced *gangway*.

caustic cracking Cracking occuring at the grain boundaries in a metal due to the combined effects of stress and corrosion in an alkaline solution. It is also known as caustic embrittlement or *stress corrosion* cracking and may occur in boilers.

cavitation This is the formation and collapse of vapour filled cavities or bubbles which occurs as a result of certain pressure variations. It can occur on propellers, pump impellers and control valve discs resulting in erosion of the material.

cavitation tunnel A closed channel in which water is circulated by an impeller in order to test model propellers. The water flow through the almost square shaped tunnel takes place in a vertical plane. See Figure C.2.

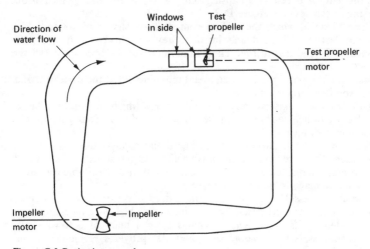

Figure C.2 Cavitation tunnel

cell The basic element of a *battery* or *accumulator* in which electrical energy is obtained as a result of chemical reactions. The electrolyte or plate material may further describe the cell, e.g. alkaline or lead-acid.

cellular A structural arrangement where a *compartment* is divided into small spaces, e.g. a double bottom.

cellular vessel A ship with cargo holds arranged for the stacking of *containers*.

Celsius scale See *temperature*.

cement box A wooden structure filled with cement and aggregate to seal a leaking pipe or valve.

cementing of well Filling the annular space between the casing and the borehole wall with cement slurry.

centistokes An established unit used for the measurement of *kinematic viscosity*. The accepted *SI unit* is m^2/s and $1 \text{ cSt} = 10^{-6} \text{ m}^2/s$. See *Redwood seconds, viscometer*.

central processing unit The *arithmetic and logic unit* and the *control unit* of a *computer* considered together. The main memory may also be considered as part of this complete unit.

centre girder A flat plate which forms a watertight longitudinal division and runs along the ship's centreline from the fore peak to the aft peak bulkhead. It may also be known as the *vertical keel* or *keelson*. See Figure F.4.

centre of buoyancy The point through which the total force of *buoyancy* is considered to act.

centre of flotation The point about which a ship changes *trim*. For small angles of trim it may be considered as the centroid of the *waterplane area* at the draught considered.

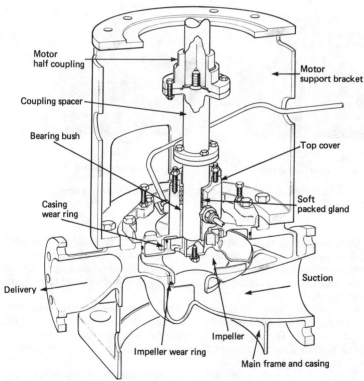

Figure C.3 Centrifugal pump

27

centre of gravity The point through which the total weight of a ship is considered to act.

centre of pressure The point at which the total pressure on an immersed surface is considered to act.

centrifugal pump A *pump* in which the liquid enters the centre, or eye, of a rotating impeller and flows radially out through the vanes. A *diffuser* or *volute* is then used to convert most of the kinetic energy in the liquid into pressure. The pump is not self-priming. See Figure C.3.

centrifuge A machine which uses centrifugal force to separate liquids or solids and liquids, see *clarifier* and *purifier*. See Figure C.4.

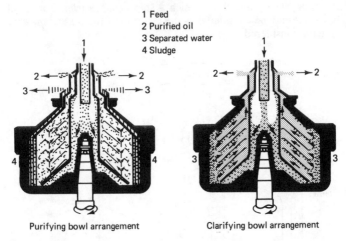

1 Feed
2 Purified oil
3 Separated water
4 Sludge

Purifying bowl arrangement   Clarifying bowl arrangement

Figure C.4 Centrifuge: (a) purifying bowl arrangement, (b) clarifying bowl arrangement

Certificate of Competency A certificate issued by a National Authority following success in an examination, a specified period of sea service and completion of other courses or requirements. Various levels of certificate exist and also designations such as marine engineer officer, third mate, master, etc.

Certificate of Registry A document issued on Government Authority following a survey. It gives details of the ship, its home port and the names of the owners. The certificate of registry establishes the nationality and ownership of a vessel. It must be endorsed with the name of the master and be in his possession.

certificated A person who holds a *Certificate of Competency*, e.g. Certificated Cook.

cetane number   A measure of the *ignition quality* of a fuel which relates to the time delay between injection and combustion. The higher the number the better the ignition quality.

chain drive   The use of a chain to drive the *camshaft*. A sprocket wheel is fitted to the crankshaft and the camshaft and an adjustable spring loaded wheel is provided for chain tightening.

chain locker   A compartment used to house the anchor chain which is located directly beneath the windlass. The opening at the top is surrounded by the spurling pipe. A perforated false floor is fitted to provided a drainage well and keep the cable dry. See Figure C.5.

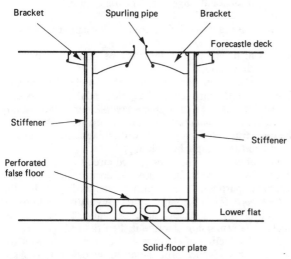

Figure C.5 Chain locker

character   Any number, letter or symbol that a computer can store or process.

character symbols   A sequence of signs, letters or numbers assigned to a classified vessel, e.g.✠100A1.

characteristics   Criteria used to describe the performance or behaviour or a system, e.g. static and dynamic characteristics of a measuring system.

charge air   A quantity of fresh air supplied to a diesel engine cylinder prior to compression.

charge air cooler   A single pass sea water cooled *heat exchanger* using finned tubes for maximum temperature reduction of the charge air. This device results in a larger mass of air entering the cylinder and also reduces the maximum cylinder pressure.

29

chart (1) A paper sheet or surface on which a permanent record is made by a recording instrument. (2) A map of the ocean or sea showing islands and coastal regions. Details of soundings, sea bed, currents, etc., are also given.

chart room A separate room or a part of the bridge where charts are stored and also used for navigation.

charter party A contract between a shipowner and a charterer for the carriage of goods either on a particular voyage (*voyage charter party*) or for a period of time (*time charter party*).

Chartered Engineer A qualification awarded by the *Engineering Council* (UK) to an engineer who has met the requirements for the award and is considered to be a professional engineer. The designatory letters are CEng.

charterer A person who hires a vessel for a particular voyage or a period of time.

check valve A valve which will permit the flow of liquid in only one direction, i.e. a *non-return valve*.

chemical cleaning The precommission cleaning of a boiler by suitable chemicals which are circulated throughout the boiler and the feed system.

chief engineer The senior engineer of a ship's complement who usually holds a Class I *Certificate of Competency* as marine engineer officer.

chill To reduce the temperature of a substance without actually freezing it.

chine A sharp edged bend in a ship's hull plating.

chip A tiny, thin piece of silicon with an integrated circuit on its surface.

chlorinating equipment A treatment plant used to ensure that water for drinking and culinary purposes meets purity standards set by the *Department of Transport* (UK) or any other regulatory authority. Sterilization is achieved by an excess dose of chlorine provided as hypochlorite tablets. The remaining chlorine is then filtered out.

chlorine injection The introduction of chlorine as a biocide to prevent the fouling of sea water systems. Hypochlorite tablets or the electrolysis of sea water is used to obtain the chlorine.

chock (1) A fitted metal block which is placed under machinery or equipment as a seating or a means of levelling. (2) A wedge used for securing purposes.

choke (1) An orifice in a pipe which controls the velocity of the flowing fluid, e.g. in the flow line of a well. (2) A coil, with a high inductance, which acts as a filter to remove alternating currents in, for example, rectifying circuits.

christmas tree An assembly or valves and fittings which is installed on the *well-head* to control the flow of high pressure oil and gas.

circuit A combination of electrical devices and conductors which, when connected together in a closed path, perform a particular function. A closed circuit is a continuous path, an open circuit is a discontinuous path. See *short circuit*.

circuit breaker A device, such as a switch or contactor, which will make or break a *circuit* under normal or fault conditions.

circuit diagram A drawing which details the functioning of a *circuit*. All the essential parts and their connections are shown using special symbols.

circulating pump A centrifugal or axial flow type *pump* which supplies large volumes of water to a system, usually for cooling purposes, e.g. jacket water circulating pump, sea water circulating pump.

cladding A coating applied to a material, e.g. a stainless steel coating bonded onto mild steel.

clampmeter A measuring instrument with tongs that are clipped around a single conductor. The current flowing can be measured without interruption.

clarifier A *centrifuge* which is arranged to separate solid impurities and small amounts of water from oil.

Class 1 fusion welding A classification society requirement for pressure vessel welding which relates to the welding plant and equipment, the quality of workmanship and periodic testing of the welding work.

class notation The *classification* details of a ship's hull or machinery using particular terms or abbreviations, e.g. oil tanker, UMS (unattended machinery space). See *character symbols*.

classification The arrangement of ships in order of merit with regard to seaworthiness by a classification society. A ship is usually classified as a result of being built according to classification rules and is then given class notation. Regular surveys are conducted by classification society surveyors to ensure that a ship is maintained according to the rules.

classification society A society which classifies, or arranges in order of merit, ships which are built according to its rules. See *Lloyd's Register of Shipping*.

cleading A covering which is used to prevent the radiation or conduction of heat, e.g. boiler casing.

clean When referring to a shipping document it is clean when not endorsed in any way which might indicate the goods are damaged.

clearance (1) A vessel has clearance when official permission is given by the authorities to leave port. (2) The distance between two closely adjoining points or surfaces.

clearance volume The volume between the cylinder head and the piston when the piston is at the end of its stroke at top dead centre.

clench pin A pin which secures the final link of the anchor chain to the ship's structure within the chain locker. Removal of this pin enables an emergency release of the anchor and chain. See Figure C.6.

clingage Oil tanker cargo which remains in the tanks after discharge. It is removed by tank cleaning and stripping.

clinometer An instrument which indicates the angle of inclination of a ship. It usually consists of a pendulum which moves through a quadrant marked in degrees.

clipper bow A bow shape where the stem has a concave form as it rises from the waterline. It is sometimes seen on passenger ships.

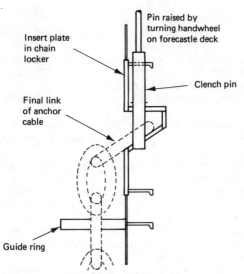

Pin raised by turning handwheel on forecastle deck

Insert plate in chain locker

Clench pin

Final link of anchor cable

Guide ring

Figure C.6 Clench pin arrangement

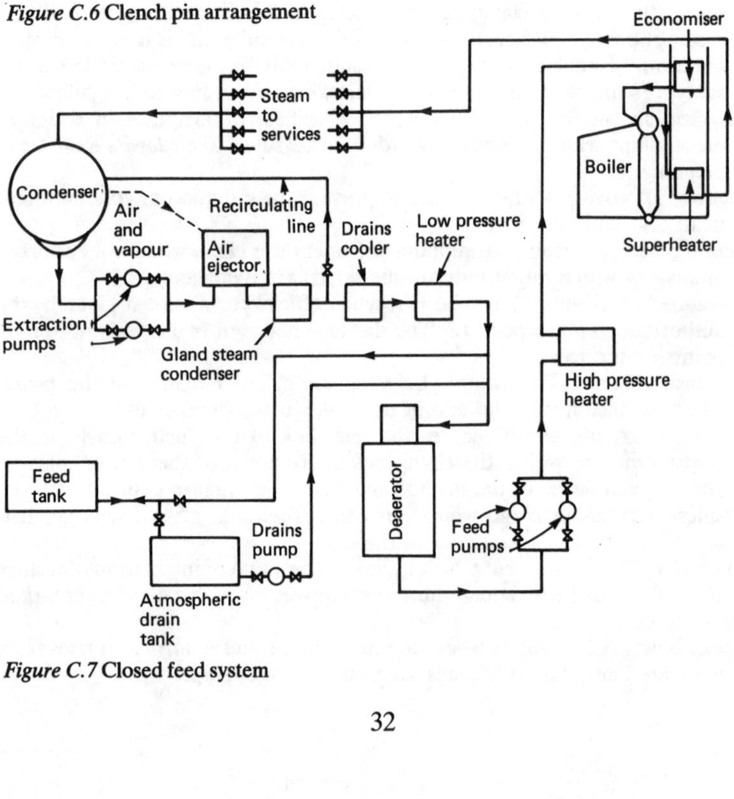

Economiser

Steam to services

Condenser

Air and vapour

Recirculating line

Air ejector

Drains cooler

Low pressure heater

Boiler

Superheater

Extraction pumps

Gland steam condenser

High pressure heater

Feed tank

Drains pump

Deaerator

Feed pumps

Atmospheric drain tank

Figure C.7 Closed feed system

clock A pulse generator which provides a common timing signal to the microprocessor, the memory and the input and output devices of a computer.

clogging indicator A display unit fitted to a filter to indicate the condition or degree of cleanliness. It is often used in hydraulic oil supply systems.

closed feed system A feed water supply system for a high pressure watertube boiler where no part of the system is open to the atmosphere. It is usual to fit deaerating equipment to remove any dissolved gases which do enter the system. See Figure C.7.

closed-loop control system A control system possessing monitoring feedback, the deviation signal formed as a result of this feedback being used to control the action of a final control element in such a way as to reduce the deviation to zero.

closing appliance Any cover or item of equipment acting as a cover for an opening in the shell, deck or a bulkhead of a ship. Classification society requirements relate to the cover, its closing and securing, and arrangements to ensure watertightness.

closure Any form of door or cover which closes any opening in the main hull, a subdivision bulkhead, a tank or any position which may affect the safety of the ship.

cloud point The temperature at which waxes form in a fuel. This can lead to pipe or filter blockage.

clutch A device which connects or separates a driving unit from the unit it drives.

coalescer A device containing a material whose surface promotes coalescence. Coalescence is the combining of small oil droplets into larger drops which will separate under the action of gravity. A coalescing filter is used downstream of an oily water separator to improve separation down to a few parts per million.

coaming The vertical plated structure around a hatchway which supports the hatchcover. The height is dictated by the *Merchant Shipping (Load line) Rules 1968*.

coaxial cable An electrical cable consisting of two conductors insulated one from another where one is in a tubular form around the other. It is used to transmit high frequency signals as it is not affected by external magnetic fields and does not produce any field.

cobalt A slightly magnetic metallic element which is used in steel alloys and for making permanent magnets.

COBOL COmmon Business Oriented Language. A high-level business oriented programming language.

cock A valve arrangement where the liquid flow takes place through a hole in a central plug. Movement of a handle attached to the plug will restrict or shut off the flow. See Figure C.8.

coefficient A factor which is used as a multiplier of the quantity or variable being considered. It is usually given a particular name, e.g. *Admiralty coefficient*.

33

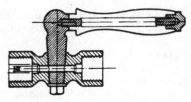

Figure C.8 Cock

cofferdam An empty space between two bulkheads or floors which prevents liquid leakage from one to the other.

coffin plate The aftermost plate of the keel which is welded to the sole piece of the *stern frame*.

coil One or more insulated conductors wound in a series of turns.

cold storage The preserving of perishable foodstuffs by storage in a refrigerated space at an appropriate temperature.

collision bulkhead The foremost major watertight *bulkhead* which extends from the bottom to the main deck.

combined framing See *framing*.

combustible material Any substance other than a non-combustible type.

combustion The burning of a combustible material resulting in the release of heat energy.

command signal An input signal to a *control system* which will determine the controlled condition value.

common rail fuel system A fuel supply system in which a single high pressure pump supplies a common manifold or rail. Timing valves determine the timing and extent of fuel delivery to the cylinder injectors.

commutator A cylindrical ring of individually insulated conductors, mounted on the rotating part of the machine, with an exposed surface upon which current carrying brushes rest.

companionway The steps leading from one deck to another.

compartment A space created when a ship is subdivided by transverse watertight bulkheads. It is a means of limiting flooding in the event of damage to the hull.

compass An instrument using either a magnet (magnetic compass) or a gyroscope (*gyro compass*) for steering or taking navigational bearings.

compensating winding A winding carrying all or part of the load current which is designed to reduce the distortion of the magnetic field by the load current.

compensation The modifying of the function or behaviour of a device or system in order to meet a specification.

compiler A program which converts *high-level programming language* into *machine language*.

complement The total number of persons that constitute the crew employed on a ship.

34

composite boiler A *firetube boiler* design which can generate steam by oil firing or the use of diesel engine exhaust gas. See Figure C.9.

compound wound The excitation of a d.c. machine which is supplied by shunt and series windings. See *flat compounded, overcompounded*.

compounding The expansion in two or more stages of the pressure or velocity change in a steam turbine. Where a high pressure and low pressure turbine expand steam one to the other this is called cross-compounding.

compression ignition engine An internal combustion engine in which the heat of the compressed air brings about combustion of the fuel injected into the cylinder.

compression ratio The ratio of the *clearance volume* and the *swept volume* to the clearance volume of a compression ignition engine cylinder.

compressor A machine which compresses a gas by reducing the volume and increasing the pressure.

computer A device which will accept data, perform prescribed functions on the data and then supply the results of the operation.

concession A licence to drill for oil or gas in a specified area.

condensate extraction pump The pump in a *closed feed system* which draws the condensed steam from the condenser, which is under vacuum, and pumps it to the deaerator.

condenser A *heat exchanger* where latent heat is removed from a gas in order to liquefy it, e.g. in a boiler closed feed system or in a refrigeration system.

condensing turbine A steam turbine which exhausts to an integral or a separate condenser.

condition monitoring The continuous measurement and recording of various parameters associated with running machinery in order to achieve optimal operating conditions and also to determine the need for maintenance in order to return the machinery to optimal running conditions.

conductance In a d.c. circuit, the reciprocal of *resistance*. In an a.c. circuit, a component part of the *admittance* which is responsible for the heat energy lost when current is flowing.

conductivity The reciprocal of resistivity. It is expressed as the ratio of current density to electric field strength and has units of siemens per metre. Conductivity measurement of water is used as a means of determining purity.

conductor A material which offers a relatively low resistance to the passage of electric current.

conduit A container for electric wires or cables to protect them from mechanical damage.

conference lines An association of shipowners in a particular trade who provide a regular service at standard rates.

connecting rod The rod which connects the *crankpin* of a reciprocating engine to the *piston* or *crosshead*. See Figure T.7.

35

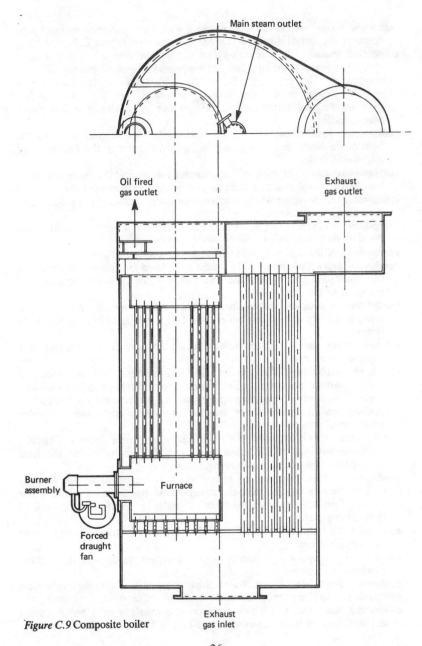

Main steam outlet

Oil fired
gas outlet

Exhaust
gas outlet

Burner
assembly

Furnace

Forced
draught
fan

Exhaust
gas inlet

Figure C.9 Composite boiler

36

Conradson carbon value A measure of the carbon residue forming ability of a fuel. It is determined by evaporation of the fuel in a closed space and measuring the percentage of carbon. The Conradson method has to a large extent been replaced by the Ramsbottom method, which gives similar results.

consignee The person to whom goods are shipped.

console A control panel, often the central unit, from which an operator can operate and supervise machinery or equipment.

constant tension winch A mooring or towing *winch* which is designed to accept a particular load. If the load or stress is exceeded then the wire will reel out (pay out), if it reduces then wire is reeled in.

constructive total loss A declaration relating to a ship or its cargo where it is damaged to the extent that the cost of recovery and making good would exceed their value.

contact feed heater A boiler feed water heater where the feed water and the heating steam are in actual physical contact, e.g. a *deaerator*.

contactor An electrical switch which is designed to frequently open and close a circuit and is not operated manually.

container A re-usable metal box of 2435 mm by 2435 mm cross-section and various lengths, e.g. 6055, 9125 or 12 190 mm. General cargo is usually carried although liquid carrying versions exist. See Figure C.10.

container freight station A cargo handling area where *containers* may be stuffed (filled) or devanned (emptied).

container ship A ship which is designed and constructed for the carriage of *containers* both in the cargo holds and on the deck. Hatch openings are the full width of the hold, cargo handling equipment is not normally fitted and high speeds up to 30 knots are usual. See Figure C.11.

continuous action Some part or a complete *control system* whose output is a continuous function of the input.

continuous maximum rated A designation for a.c. and d.c. motors and generators which indicates that they are designed for continuous operation at a particular load.

continuous sea service rating The power output of an engine which will be obtained during normal sea service conditions on a continuous basis. See *rating*.

continuous survey An alternative to periodical *surveys* where all the compartments of the hull or parts of the machinery that require a periodic survey are examined in rotation with an interval of five years between consecutive examinations of each part. See *classification*.

contra rotating propellers A propulsion arrangement with two propellers rotating in opposite directions on the same shaft.

control action The relationship between the input and output signals of a *control system*.

control media Any medium used for the transmission of signals and actuation of equipment, e.g. compressed air, hydraulic oil and electricity.

control stations The spaces in which are housed the radio, navigation

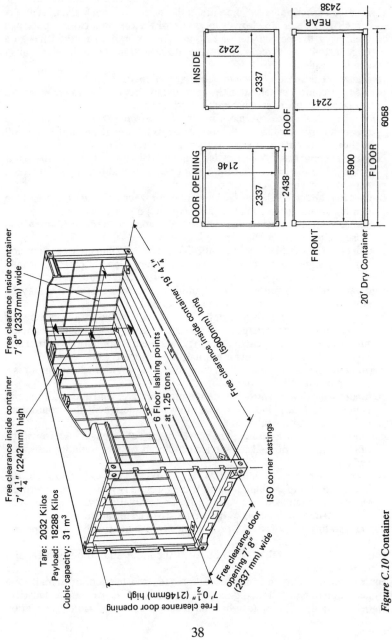

Tare: 2032 Kilos
Payload: 18288 Kilos
Cubic capacity: 31 m³

Free clearance inside container
7' 4 1/4" (2242mm) high

Free clearance inside container
7' 8" (2337mm) wide

6 Floor lashing points
at 1.25 tons

Free clearance inside container 19' 4 1/4" (5900mm) long

ISO corner castings

Free clearance door opening
7' 0 1/2" (2148mm) high

Free clearance door opening 7' 8" (2337 mm) wide

INSIDE

2242

2337

DOOR OPENING

2146

2337

REAR

2438

ROOF

2241

5900

FLOOR

2438

FRONT

6058

20' Dry Container

Figure C.10 Container

38

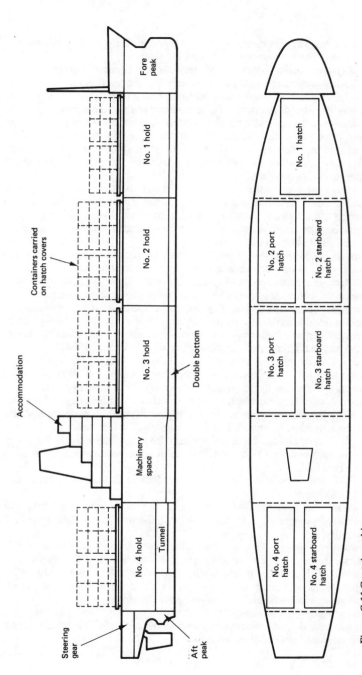

Figure C.11 Container ship

equipment, the emergency generator and the fire recording equipment. They must be surrounded by *A-class divisions.*

control system An arrangement of elements interconnected and interacting in such a way as to maintain, or to affect in a prescribed manner, some condition of a body, process or machine which forms part of the system.

control unit This device, as part of a *computer* enables the *central processing unit* to execute a particular task by orders to units inside and outside of it.

control valve A valve which regulates the flow of a fluid. It is usually operated remotely as the correcting unit of an automatic control system. The actuator or motor unit may be pneumatic, electric or hydraulic in operation. See *pneumatic control valve.*

controllable pitch propeller A propeller made up of a boss with separate blades mounted into it. An internal mechanism enables the blades to be moved simultaneously through an arc to change the pitch angle and therefore the pitch. Astern thrust can be produced without changing the direction of rotation of the propeller.

controlled condition The physical quantity or condition of the controlled body, process or machine which the system is to control.

controller In a process control system this unit will combine the function of the input, comparing, amplifying and signal processing elements.

Convention The formal assembly for the consideration of legislation or important matters. In a marine context it is a special meeting of the *International Maritime Organization* which produces a particular text, e.g. International Convention on the Safety of Life at Sea, 1974.

cooler A *heat exchanger* arranged to remove heat from a flowing fluid.

Cope's regulator A feedwater regulator for a small, low pressure boiler which operates on the measured water level, i.e. single element.

copper A ductile metal which has good electrical conductivity and is much used in electrical equipment. It has a high resistance to corrosion and also forms a number of important alloys, e.g. brass and bronze.

core (1) A magnetic material upon which the magnetizing coil is fixed, e.g. a transformer core. (2) A conductor and its insulation, which is part of an electric cable.

correcting unit The motor and correcting element in a control system. It is often a *control valve.*

correction An amount which must be added to or subtracted from the indicated value of an instrument to obtain the true value of the measured quantity.

corresponding speed The speed at which two geometrically similar ship forms are run at when they conform to *Froude's Law of Comparison.*

corrosion The chemical or electrochemical reaction of a metal or an alloy with its environment which results in its deterioration. See *crevice corrosion, electrochemical corrosion, galvanic corrosion, rust.*

corrosion piece A short length of steel or iron pipe fitted into a copper alloy or galvanized piping system to act as a sacrifical anode. It is fitted where it can be easily removed and replaced.

40

corrugated bulkhead The use of corrugations or swedges in a plate instead of stiffeners which produces as strong a structure with a reduction in weight. The troughs are vertical on transverse bulkheads but horizontal on longitudinal bulkheads.

cost and freight A price quoted for goods which includes their cost and freight charges.

cost and insurance A price quoted for goods which includes their cost and insurance charges.

cost insurance freight A price quoted for goods which includes their cost, insurance and freight.

counter (1) The part of the *stern* which overhangs the *rudder*. (2) The instrument which indicates the number of revolutions of the propeller. (3) A circuit which counts electronic pulses, e.g. digital counter.

coupling (1) An electrical linkage between two circuits to enable the transfer of energy. (2) A fitting between two shafts to enable the transfer of a rotary motion. (3) The interdependence of, for example, the motions of a ship in the water. See *cross coupling*.

covered electrode A metal electrode, used in electric *arc welding*, which has a coating which assists the welding process.

cowl The shaped top of a natural ventilation trunk which may be rotated to draw air into or out of the ventilated space.

crack arrester A band of tough steel which is used as part of the hull structure to prevent a fracture from spreading.

crack tests Any non-destructive test to examine a material for cracks. See *dye penetrant testing, magnetic crack detection*.

cracking A general term for the conversion process which produces lighter oils from heavy oils. The main types are *thermal cracking, catalytic cracking* and *hydro-cracking*.

cradle A supporting framework used for launching a ship. It consists of a fore-poppet structure and an after-poppet structure both of which move with the ship as it is launched.

craft A ship or boat of any type or size.

crank An arm or link connected to a shaft to enable a transfer of motion.

crank angle The angle of a crank of a crankshaft usually given by reference to top or bottom dead centre.

crank pin The pin fitted between the webs of a crankshaft and to which the big end of the connecting rod is attached.

crank throw The radial distance from the centre line of a crankshaft to the centre of a crankpin. It is equal to half of the stroke.

crank web The arm or side of a crank.

crankcase The casing which encloses the crankshaft and the lower end of the connecting rod. In some engines it is used as an oil storage tank or sump. See Figure T.7.

crankcase monitoring The use of sensors to measure or detect the presence of oil mist, high temperature or flammable vapours and continuously measure or indicate the value. An alarm will be given if set limits are

41

exceeded and conditions likely to cause an explosion exist.

crankshaft The main driving or driven shaft of a reciprocating engine which has one or more cranks.

crankshaft deflection A measurement taken by a micrometer dial gauge, inserted between the crank webs, as the crankshaft is moved through one revolution. It is a method of checking engine alignment.

creep The slow plastic deformation of a material under a constant stress. It is a particular problem at high temperatures e.g. in boiler superheater tubes.

crevice corrosion A form of galvanic action which occurs when a crevice is formed between two surfaces and the inside is anodic with respect to the outside surface.

crew The complement of a ship with the exception of the Master. It is sometimes used to refer to the complement with the exception of all officers.

Criterion of Service Numeral A number which is used in subdivision calculations. It is based on the relationship between volumes for passengers and machinery and the total volume of the vessel. See *factor of subdivision*.

critical path method A *network analysis* technique which is used to graphically represent a programme of events from the beginning to the end. The critical path is the route which takes the longest time.

critical pressure The pressure at which a gas will just liquefy at its critical temperature.

critical speed A speed at which a rotating shaft or system becomes dynamically unstable. It occurs due to *resonance* between pulsations in the engine torque and a natural frequency of lateral vibration of the shaft system. See *torsional vibration*.

critical temperature The temperature above which a particular gas cannot be liquefied. It is important for *refrigerants* which must have a high critical temperature.

critically damped The minimum degree of *damping* in an instrument or control system which will prevent oscillation after an abrupt change.

cross coupling With reference to a ship moving in a seaway this is where one motion, e.g. heaving, brings about forces or moments which create another motion, e.g. pitching.

cross curves of stability Graphs of righting arm, *GZ*, to a base of displacement for various constant angles of heel. They are used to determine the extent of a vessel's stability at any particular displacement. See Figure C.12.

cross flooding Where a compartment on one side of a ship is flooding and it is beginning to heel excessively, the corresponding compartment on the opposite side may be flooded to eliminate the angle of heel.

cross ties Horizontal stiffening structures which are fitted in the wing tanks of oil tankers between the side shell and the longitudinal bulkhead. See Figure F.4.

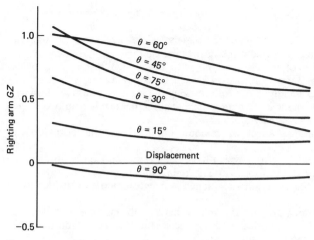

Figure C.12 Cross curves of stability

cross trees Horizontal members positioned across the upper part of a mast to enable derricks to be supported away from the mast.

crosshead A reciprocating block which usually slides in guides and is the connecting point for the piston rod and the connecting rod in a slow speed, two-stroke diesel engine or for the rams and the tiller in a steering gear.

crown block The lifting tackle which is fitted at the top of a derrick to support the drill string. It consists of several sheaves which are mounted on a shaft in a rectangular frame. See Figure D.2.

crude oil A liquid mixture consisting mainly of different hydrocarbons. It is the raw material from which other petroleum products are obtained.

crude oil washing A cargo oil tank cleaning system in which crude oil is used as a spray to remove sludge from the surfaces.

cruise liner A luxury *passenger ship* which sails to specific ports on a set schedule.

cruiser stern A fully radiused stern construction. Cant frames and *cant beams* are used in this well-stiffened structure.

crutch A supporting structure for a derrick or crane jib comprising a post and a shaped support piece.

cupro-nickel An alloy of *copper* and *nickel* with 20 to 30% of nickel. It has good strength properties and resistance to corrosion. It is often used for condenser tubes.

current The flow of electricity along a conductor. The unit of measurement is the *ampere*.

current transformer An *instrument transformer* which uses the current transforming property of a transformer. The secondary winding is connected to the measuring instrument.

cursor A small flashing line or rectangle which is displayed on a computer monitor to indicate where the next typed character will appear.

Curtis turbine An impulse turbine with more than one row of blades following each row of nozzles, i.e. velocity compounding.

curve of statical stability A graph showing the righting arm, *GZ*, against a base of angle of inclination for a fixed displacement. It is readily obtained from a set of *cross curves of stability*.

cut A particular product or fraction obtained by the distillation of *petroleum*.

cuttings (1) Pieces of rock which have been removed from a borehole by the action of the drill bit. (2) The removable parts of the bulwarks or guard rails which permit the gangway or accommodation ladder to be entered.

cutwater The forward part of a ship which cuts through the water.

cybernetics The theoretical and practical study of communication and control in living organisms or machines.

cycle A recurring sequence of events or values during a period of time. In an internal combustion engine the sequence of events are induction, compression, power and exhaust. A *two-stroke* or a *four-stroke cycle* may be used.

cylinder A cylindrical cross-section chamber in which a piston may move freely.

cylinder block A metal casting forming part of a reciprocating engine which contains one or more cylinders.

cylinder head A metal casting which fits over one or more cylinders of a cylinder block. A cylinder head gasket ensures a gas tight fit between the two when they are bolted together. See Figure T.7.

cylinder liner A replaceable cylinder which fits into a cylinder block.

D

damp To reduce the amplitude of a movement such as a vibration or an oscillation.

damper (1) A hinged flap used to control gas flow, e.g. in a boiler uptake. (2) An energy dissipating device which reduces the amplitude of certain vibrations.

damping The process which dissipates the energy of a vibrating system in order to reduce the amplitude of the vibration or oscillation.

dangerous goods Any goods which are so classified in any acts, rules or by-laws or that have similar characteristics, properties or hazards. The *International Maritime Organization* has produced a code for the carriage of dangerous goods in ships (IMDG Code). This code recognizes nine broad classes of such goods and is universally accepted. Requirements for their carriage are also made in the International Convention for the Safety of Life at Sea, 1974.

dangerous spaces Any compartment or space in a tanker where flammable or explosive vapour might accumulate. Examples would be cargo tanks, the cofferdams adjoining cargo tanks and cargo pump rooms.

dash pot A damping device comprised of a piston in a cylinder. Liquid or air may be used to provide the fluid friction.

data buoy An unattended floating platform which acts as a hydrographic station to obtain and transmit weather information.

data logger A printer used as part of a control system to provide a log or record of parameters either on demand or at set intervals.

database A comprehensive collection of information which can be obtained using a suitable computer program in a variety of forms in order to meet specific needs, e.g. names in alphabetical order.

datum A reference point or line from which measurements are made.

davit A support arm which acts as a lifting arrangement. It may be used as a crane to lift stores or as a support for lifeboats. See *gravity davit*.

dead band The region in which a change of input signal to a unit will cause no change of the output signal.

dead circuit An electric circuit which has no applied voltage nor any electrical charge. A circuit which is safe to work upon.

dead slow The slowest speed at which a ship can still be manoeuvred.

deadfreight The freight paid for cargo space which has been booked but not used.

deadlight A hinged steel cover which is part of a port or scuttle. It is used to cover the glass in heavy weather.

deadrise Another term for *rise of floor*.

deadweight The difference between the *displacement* and the *lightweight* of a ship at any given draught. It is the mass of cargo, fuel, stores, etc., that a ship can carry and is measured in tonnes.

deadwood The flat, vertical portion of the after end of a ship. It has an influence on the turning of a ship since if it is reduced the turn is easier.

deaerator A direct contact feed heater in which the feed water and steam mix. The heating and mixing promotes the release of dissolved gases which can cause corrosion problems in a boiler.

decibel A unit which is one-tenth of a bel. It is used to compare levels of electrical power or sound.

deck The horizontal surface which completes the enclosure of some part of the hull.

deck beam A stiffener located beneath a deck and running transversely across a ship. See Figure F.4.

deck line This is a horizontal line 300 mm long and 25 mm wide which is positioned amidships port and starboard, see Figure L.5. The upper edge of the line is located level with the upper surface of the freeboard deck plating on the outer shell. See *load lines*.

deck seal A non-return valve arrangement to prevent the back-flow of flammable gases, from the cargo tanks, into an *inert gas plant*.

deckhead The underside of a deck.

deckhouse An item of *superstructure* which does not extend the full breadth of a ship.

Declaration of Ownership A statement which must be made by any person who wishes to become the owner of a British ship.

decompression sickness See *bends*.

deep tank A tank which extends from the shell or double bottom up to or beyond the lowest deck. It is usually arranged for the carriage of fuel oil or water ballast but may be fitted with hatches and used for cargo.

deepwell pump A multi-stage *centrifugal pump* which is driven from the deck and provides a high discharge head. It is usually located at the bottom of a cargo tank and discharges only from that tank. See Figure D.1.

degree API An oil classification system based on density and defined by the *American Petroleum Institute*. The numbers range from −1 to +101 with larger numbers indicating lighter oils, e.g. heavy crude 20 API, light crude 40 API.

dehumidifier A substance or a unit which removes moisture from the air in a space or system. *Calcium chloride* and *silica gel* are water absorbing chemicals (*desiccants*) which may be used.

dehydration The removal of water from a substance, e.g. crystals, oil, or a refrigerant, by chemical action, heating or distillation.

Delaval turbine A high speed *impulse turbine* which has only one row of nozzles and one row of blades.

delivered power The propulsive power which must be available at the propeller in order to drive a particular design of ship at a given speed.

demise A ship chartering term similar to bareboat. The shipowner charters out his vessel to another party who assumes total control for the period of charter.

46

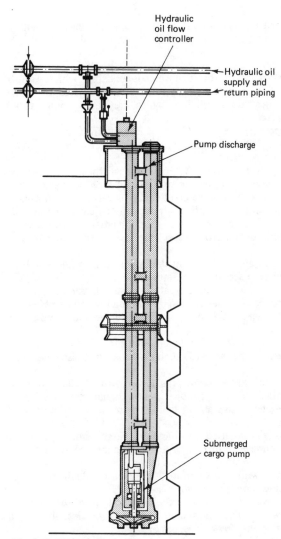

Figure D.1 Deepwell pump

demodulator A device used to recover a transmitted signal from a modulated *carrier wave*.

demulsibility The ability of an oil to mix with water and then release the water in a *centrifuge*. This property is also related to the tendency to form *sludge*.

47

demurrage A compensation payment to a shipowner when a ship is delayed, during loading or discharge, beyond the period agreed by the charterer.

Department of Transport The current name of the UK Government Department, the Marine Directorate of which is concerned with the safety of ships and crew. Surveyors are employed to examine vessels to ensure compliance with rules and regulations which are based upon various *Merchant Shipping Acts*.

Department of Transport Enquiry A Court of Enquiry into an accident or casualty involving a British ship, the passengers or crew, in particular, where loss of life has occurred.

Department of Transport Surveyor A Government official who surveys ships and their equipment to ensure compliance with statutory requirements. He will also examine candidates for various certificates.

depletion A reduction in the contents of an oil well or field which may be physical, e.g. the well is empty, economic, e.g. the costs exceed the value of the oil obtained, or natural, where the gas drive mechanism is insufficient.

depth moulded The depth of a ship from the upper deck to the base line, measured at the midship section. See Figure P.7.

derating The operation of a diesel engine at the normal maximum cylinder pressure for its *continuous sea service rating,* but at a lower mean effective pressure and shaft speed. It is a means of reducing the *specific fuel consumption.*

derivative action The action of a control element where the output signal is proportional to the rate of change of the input signal. It is also called *rate action.*

derivative action time In a proportional plus derivative controller this is the time interval in which the proportional action signal increases by an amount equal to the derivative action signal, when the rate of change of deviation is constant.

derrick (1) A lifting device used for handling cargo, stores or heavy equipment. (2) A framework forming a tower over the drilling slot. It is built onto the deck of a drilling rig. See Figure D.2.

derrick barge A barge with one or more high capacity, high lift cranes which can lift modules onto production platforms.

desalination The removal of the various chemical salts from sea water to produce distilled water. The equipment used may be described as a *distiller,* an *evaporator* or a *fresh water generator.*

desiccant A substance which will absorb moisture, e.g. anhydrous calcium chloride. It is used as a drying agent

design spiral A design procedure which begins with certain basic data and a cycle is undertaken which spirals out from the centre. As the cycle is repeated fewer designs are considered but in greater detail until a final optimal design is obtained. See Figure D.3.

desired value The value of a controlled condition which the system is

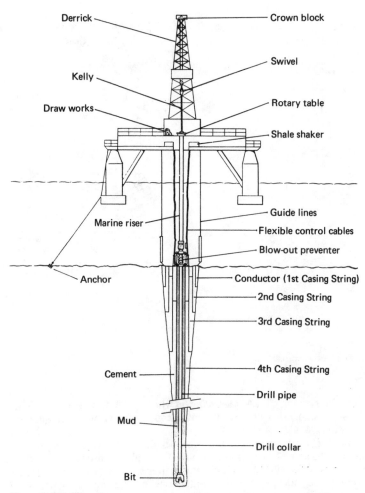

Derrick — Crown block
Kelly — Swivel
Draw works — Rotary table
— Shale shaker
Marine riser — Guide lines
— Flexible control cables
— Blow-out preventer
Anchor — Conductor (1st Casing String)
— 2nd Casing String
— 3rd Casing String
— 4th Casing String
Cement — Drill pipe
Mud —
— Drill collar
Bit —

Figure D.2 Drilling rig

required to maintain.

despatch days The days saved by loading or discharging quicker than stated in the charter party. The charterer may seek compensation if a provision exists to this effect.

desuperheater Another name for an *attemperator*.

detecting element The item in a measuring or control system which responds directly to the value of the *controlled condition*.

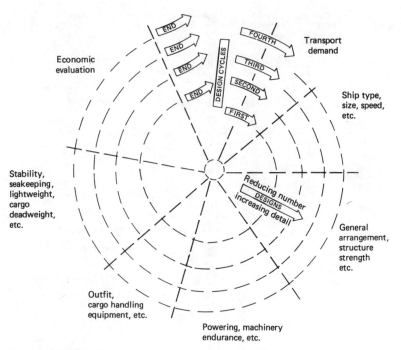

Figure D.3 Design spiral

detergent lubricating oil A *lubricating oil* which will hold solid contaminants in suspension. It will also inhibit lacquer formation, corrosion and oxidation.

detonation (1) A form of explosion where high velocities and high pressures occur and a shock wave travels through the explosive mixture. A hammer blow or 'knock' will occur which may shatter any enclosing structure, e.g. a crankcase door. (2) The spontaneous combustion of part of the compressed mixture in a petrol engine after the spark has occurred. There is an accompanying 'knock' or noise. In a mild form it is called pinking.

detuner An auxiliary vibrating or rotating mass connected by springs to a system in order to modify its vibration characteristics.

devanning The unloading of a *container*.

devaporizer A vent condenser through which the gases released by the *deaerator* pass in order to give up their heat energy to the circulating feed water.

developed area The area of a curved surface when laid out on a flat surface, e.g. a propeller blade. See Figure P.8.

development well A well, drilled from a production platform, to increase production from a field or to provide for the injection of a drive mechanism such as water or gas.

deviated drilling A drilling technique where the hole gradually veers away from the vertical in order to enter the well at a set distance from the main vertical well.

deviation The difference between the measured value of the *controlled condition* and the *command signal*.

devils claw A stretching screw with two heavy hooks or claws. It is used to secure the anchor in the *hawse pipe*.

dew point The temperature at which water vapour in the atmosphere precipitates as drops of water or condensation on surfaces. At this temperature the *vapour pressure* and the *saturation* vapour pressure of the water in the air are equal.

diac A semiconductor device comprised of two *silicon controlled rectifiers* connected back-to-back on a single piece of silicon but without the gate terminal. It may be used in the triggering circuit of a *triac*.

diamagnetic A material with a relative *permeability* less than unity.

diaphragm (1) A circular plate made up of two halves. It fits between the rotor wheels of a steam turbine and has nozzles in its upper half. (2) A flat or corrugated thin metal plate used in pressure measuring instruments. (3) A piece of flexible synthetic rubber material which forms part of a pressure-tight chamber for a pneumatically operated valve actuator.

dielectric A solid, liquid or a gas in which an electric field can be maintained with little or no external electrical energy supplied. It is therefore an *insulator*.

diesel engine An internal combustion reciprocating engine in which the fuel is ignited, after injection, by the hot compressed air present in the cylinders. An oil or gas fuel may be used.

diesel index number A number which indicates the *ignition quality* of a fuel. The higher the value the better the fuel.

$$\text{Diesel index number} = \frac{G \times A}{100}$$

where G = gravity obtained at 60°F by a hydrometer having the API scale. A = aniline point in °F; this is the lowest temperature at which equal parts by volume of the fuel oil and freshly distilled aniline are completely miscible.

differential pressure The difference in pressure existing between two points.

diffuser A chamber, which surrounds the impeller of a centrifugal pump or a compressor, in which some of the kinetic energy of the fluid is converted into pressure energy due to an increasing cross-sectional area of the flow path.

digital computer A *computer* which uses discrete signals to represent numerical values.

dimensional analysis The use of the dimensions of physical quantities in order to determine their interdependence. For example, if a quantity Q depends upon a, b, c, d, then Q must be some function of these, i.e. $Q = F(a, b, c, d)$. Also a, b, c, d must be grouped together so that they give the correct dimensions for Q.

DIN rating The German industrial standard (Deutche Industrie-Normen).

diode A semiconductor device formed by a junction of n-type and p-type materials. It effectively permits current flow in only one direction.

dionic unit An electrical *conductivity* measurement using a unit of microsiemen/cm^3, corrected to 20°C.

direct bilge suction A direct suction pipe from the machinery space bilge which is connected to the largest capacity pump. It is also known as the *emergency bilge suction* or the *bilge injection valve*.

direct current An electric current which flows in only one direction and is virtually free of pulsation or changes in value.

direct digital control A form of automatic control where a computer provides outputs which directly control a process.

direct drive A propulsion system arrangement where the engine crankshaft is directly coupled to the propeller shaft. It is usual with *slow speed diesel* engines.

direct expansion refrigeration A *refrigeration* system in which the liquid refrigerant is expanded by passing through a regulating valve.

direct injection A combustion chamber arrangement in a four-stroke diesel engine where the fuel is injected into the combustion chamber formed by the piston crown and the cylinder head.

direct on-line starting A starting arrangement for an *induction motor* which results in a large starting current.

direction finder A radio receiver with a loop aerial which can be rotated to determine the bearing of a transmitting station or radio beacon.

directional drilling See *deviated drilling*.

disbursements All payments made for a ship's expenses, e.g. fuel, stores, port charges, etc.

disc The movable part of a *valve* which provides a variable restriction to fluid flow.

disc area ratio The developed blade area of a propeller, including the boss, divided by the area of the circle where the diameter, D, is the propeller diameter.

discharge (1) To unload cargo. (2) Any liquid which is pumped overboard after use.

discharge book An identity document issued to officers and seaman. It contains details of the holder, previous service and a conduct report.

discontinuity Any break or change in section, thickness or amount of plating material in any part of a ship's structure.

discontinuous action Some part, or a complete control system, whose output is a discontinuous function of the input. Examples are *on–off* and bang–bang.

discovery well An *exploration* well which has been successful in finding hydrocarbon deposits.

discrimination The smallest change in the measured quantity which will produce an observable movement of the index or pointer.

disembark To leave a ship.

disengaging gear Equipment to enable the rapid release of a *lifeboat* from the 'falls' or wires with which it is lowered to the water.

disk A flat circular metal or plastic disk with magnetic surfaces on which a computer can write and therefore store data.

dispatch money A bonus payment to a charterer for loading or discharging a vessel in less time than stated in the *charter party*.

displacement The mass of water displaced by a floating ship, measured in tonnes.

displacement pump A pump whose operating action is achieved by the reduction or increase in volume of a space by a mechanical action which physically moves the liquid or gas.

dissolved gases Any gas which is present in water and may be released when it is boiled. The principal gases are oxygen and carbon dioxide which must be removed from boiler feed water to prevent corrosion.

dissolved solids Any impurities present in pure water either produced by an evaporator or used as boiler feed water. Units used are parts per million (p.p.m.) where one p.p.m. represents one part of solid matter in one million parts of water by mass.

distillate Any product obtained by distilling *petroleum* and condensing the vapours. In marine practice it refers to any liquid fuel in the viscosity range 30–50 seconds Redwood No. 1 at 37°C.

distillation A process in which a liquid is converted into a vapour which is then condensed to produce a liquid distillate. It is part of the refining process for crude oil and is also a means of producing fresh water from sea water.

distiller Equipment used to convert sea water into distilled water suitable for boiler feed water or, after further treatment, drinking water. A boiling or a 'flash' process may be used. Other names used for this equipment include *evaporator* and *fresh water generator*.

distortion (1) In shipbuilding this refers to the effects of welding on a metal plate. It may appear as shrinkage or an angular twisting. (2) In signal transmission it is any change from a specified input–output relationship in a range of amplitudes, frequencies, phase shifts or during a time interval.

distribution (1) The provision of an electricity supply to various items of equipment, often at different voltages. (2) The manner in which a particular measured quantity is spread over some record, e.g. Rayleigh distribution of actual wave heights.

distribution board A grouping of electrical equipment in a distribution system, which supplies minor items such as lighting and is itself supplied from a section board.

disturbance Any change inside or outside a *control system* which upsets the equilibrium.

53

diverter   A type of *blowout preventer* used on an oil rig to safely divert fluid from a well under 'kick' or sudden pressure rise conditions.

dock   A place where a ship can be moored or the action of mooring. A dry or graving dock can be pumped dry to enable hull repairs and maintenance. A wet dock is a port area which is isolated from tidal water movements by a lock gate. A floating dock can be submerged in order that a ship may enter and is then raised to lift the ship out of the water for repairs and maintenance.

dock dues   Payments made for the use of a dock and its facilities.

docking bracket   A vertical stiffener fitted between each transverse to support the centreline girder of an oil tanker.

docking plan   A longitudinal section drawing of a ship showing the transverse bulkheads, machinery space and sea connections. It also details cross-sections at various points showing the form of the bottom. It is used to prepare a dry dock with appropriate blocking to receive and support the ship.

docking plug   A plug consisting of a threaded bolt which is fitted in all double bottom tanks to enable draining prior to examination in dry dock.

docking stresses   The complex stresses occuring in a ship's structure when it is not supported by water pressure, e.g. in a dry or floating dock.

docking survey   The survey of a ship in a dry dock or on a slipway. It is a *classification society* requirement at set intervals according to the age of the ship.

dog   A small metal fastener or clip used to secure doors, hatchcovers, etc.

dog shores   Supporting timbers for a ship being built on a slipway. They are located between the standing, or fixed, ways and blocks on the sliding way and hold the ship until launching when they are released.

doghouse   (1) A shelter on the drilling floor of an oil rig which acts as an office or base for the driller. (2) A compression and decompression chamber in a diving installation.

dogleg   A well or *borehole* in which there is a sharp bend or abrupt change in direction.

domestic water   Water which may be used for drinking or cooking. It will have been suitably treated to kill off bacteria normally by chlorination.

donkey boiler   An auxiliary *firetube boiler* on a motor ship.

dot matrix   A dot pattern, e.g. seven dots high by five dots wide, which is used by a particular type of *printer* to produce written data. A print head actuates wires or needles from the matrix to print any particular letter or number.

double-acting   An operational event which takes place on each side of the piston, e.g. in an internal combustion engine or pump.

double bottom   A structural arrangement which uses an inner and outer bottom shell with supporting flooring in between. It provides a degree of protection should the outer shell be damaged and the tank spaces so formed are used for the storage of various liquids.

double evaporation boiler   A *boiler* with two independent systems for steam

54

generation in order to avoid contamination of the feed water. The primary circuit is a conventional **watertube boiler** which provides steam to the heating coils of a **steam-to-steam generator,** which is the secondary system.

double insulation The use of two layers of insulation on portable electrical equipment which has accessible metal parts.

double reduction gearing A compact arrangement of helical gear wheels in which two stages of speed reduction are used. The high speed input drives a primary pinion which engages with a primary wheel. The primary wheel drives, on the same shaft, a secondary pinion which engages with the main wheel. See Figure D.4.

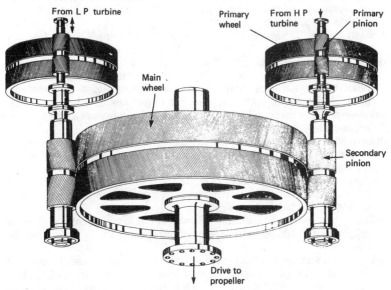

Figure D.4 Double reduction gearing

dowel A fitted circular pin which is used to accurately position parts of a machine, e.g. a cover.

down to her marks A vessel is loaded to the maximum permitted draught for the relevant load line. See **load lines.**

downcomers Large bore pipes which are fitted between the steam and water drums of a boiler and pass outside of the furnace. They ensure a natural downward flow of cooler water during steam generation.

downtime The period when a drilling rig or other large item of equipment is unable to operate due to bad weather or other problems.

Doxford engine A single acting two-stroke slow speed opposed piston engine which, although once very common, is no longer in production.

drag (1) The forces set up in the water which oppose the motion of a ship. (2) A component force acting on an aerofoil shape which is moving relative to a fluid, e.g. a rudder.

drag chains Large heavy bundles of anchor chain which are secured by the wire ropes to the hull of a ship to slow down or direct the motion of a ship during launching.

drain hat A bilge water collecting point in a continuous tank top.

drains cooler A *shell and tube heat exchanger* which receives drains steam from various auxiliaries and condenses it. The heat energy from the steam is received by circulating feedwater.

draught (1) The distance from the waterline to the keel. It is called extreme draught when measured to the underside of the keel. See Figure P.7. (2) The flow of air through a boiler furnace.

draught marks (1) Figures cut into the stem and stern. (2) The *load line marks*.

draw card An engine crank angle against pressure diagram usually taken by hand using an *engine indicator*. The indicator drum is 90 degrees out of phase with piston stroke and the term *out of phase diagram* is sometimes used. Its shape may enable the detection of fuel timing or injector faults.

draw-down The pressure drop which occurs at the bottom of a well between a static and a flow condition.

draw-works The complete winch mechanism and the crown and travelling blocks which are used for hoisting and lowering drill pipe, casing and tubing. See Figure D.2.

dredger A vessel which excavates the sea or river bed in order to provide a deeper channel or dock. Various types exist which use buckets, grabs or suction pumps to lift the material.

dribble The leakage of fuel from injectors at too low a pressure to atomize properly. It results in exhaust smoke and is caused by a leaking needle valve.

drier (1) A unit used to remove moisture from instrument air. It may use a *desiccant* or a small refrigeration plant operating as an after cooler. (2) A unit used in a refrigeration plant to remove moisture from the system.

drift (1) The distance and direction moved by a ship at sea as a result of wind or current. (2) The horizontal distance between a perpendicular and the position of a deviated well, at some given depth.

drill collar Lengths of very heavy pipe which are fitted around the drill pipe directly over the drill bit. They exert a pressure on the drill bit and also align it to ensure a straight true hole.

drill pipe The pipe which carries and rotates the drill bit. The drilling fluid passes down the inside of the pipe. See Figure D.2.

drill ship A mobile, self-propelled, drilling rig with the hull shape of a ship which is designed for operation in deep water. Such vessels are usually fitted with *dynamic positioning* equipment in order to maintain an accurate position when drilling. See Figure D.5.

drill stem testing A test to determine the productive capacity of a well. The

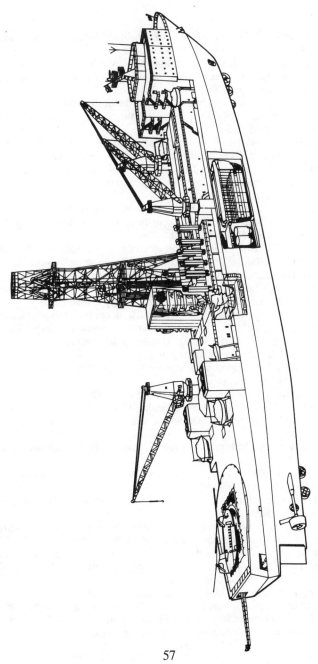

Figure D.5 Drill ship

well contents are allowed to flow into the drilled hole or drill pipe and out through a test tree with valves, chokes, etc.

drill string The complete drilling assembly which is suspended from a swivel in the derrick. It comprises the kelly, drill pipe, tools, collars and the drill bit. See Figure D.2.

drilling fluid The liquid which is circulated down the drill pipe through the drill bit and back up the well. It is a mixture of clay suspended in water and is often called '*mud*'.

drilling platform The structure which provides a flat surface, above the sea, upon which the drilling rig and other facilities are mounted. A platform may be described as fixed or floating or in terms of its use as exploration, development or production or with respect to the method of construction, e.g. steel, concrete or hybrid.

drilling rig The complete assembly of all machinery and the supporting structure from which a well can be drilled. See *drill ship, jack-up drilling unit, semi-submersible drilling unit*.

drip-proof A form of *enclosure* for electrical equipment which provides protection from falling liquid or liquids being drawn in by ventilating air.

droop (1) The change in engine speed following a load change. It is usually about 4% for a full load change and can be varied in the governor. (2) Another word for offset in a proportional-only controller.

dry bulb temperature A measurement which, in conjunction with a *wet bulb temperature,* enables the determination of *relative humidity*. The unit containing the two thermometers is called a *psychrometer*.

dry dock A dock with a watertight gate so that vessels may float in and then the water can be pumped out. Repairs and inspections of the underwater hull can then take place.

dry hole A *borehole* which did not contain any hydrocarbon.

dry powder The contents of a particular type of fire extinguisher. The powder is sodium bicarbonate which will smother the fire and exclude oxygen.

dry sump A *sump* which does not store liquid oil. It is drawn out or drains to a separate tank.

dual completion An arrangement within a well to enable production from two reservoirs simultaneously.

dual fuel An engine or boiler which has been designed to use oil or gas or a mixture as its fuel.

duct keel An internal passageway, formed by twin longitudinal girders, which extends a considerable distance along the length of the ship. Piping for the various holds and tanks is run along this passage. See Figure D.6.

ductility The ability of a material to undergo permanent change in shape without rupture or loss of strength.

dumb barge A barge which has no means of propulsion.

dummy piston A piston which is mounted on the end of a reaction turbine rotor to balance out axial thrust. Steam is admitted to a cylinder assembly around the piston to produce the balancing thrust.

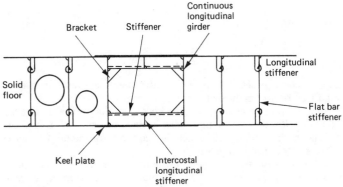

Figure D.6 Duct keel

duplex filter The use of two filters in parallel, usually with a simple valve changeover arrangement. It is used in fuel or lubricating oil systems to enable rapid changeover without interruption of flow.

duplexing The simultaneous operation of a communications channel in both directions, e.g. in radar for the transmission of a pulse and the receiving of the echo.

duty engineer The engineer responsible for machinery and equipment operation during *unattended machinery space* operation.

dye penetrant testing A non-destructive test appled to metal items to detect surface cracks. A penetrant liquid is applied which is then detected within cracks by the use of a developer.

dynamic error The difference between the ideal and actual response of a control or measuring system after a change.

dynamic positioning A method of accurately locating a floating drill ship or rig over a well. Position may be determined from instruments on the sea bed or by satellite signals. Movement is usually achieved by computer-controlled *thrusters*.

dynamic stability The ability of a controlled system to return to a stable state after a disturbance.

dynamical similarity An equivalence between two systems, such as a ship and a model, when the magnitude of forces at similar points are in a fixed ratio at corresponding times.

dynamical stability The work which must be done upon a ship to heel it to a particular angle.

dynamically supported craft A vessel supported other than by the hydrostatic or buoyancy force of water which is displaced by the hull. See *hovercraft, hydrofoil.*

dynamo An electrical machine which converts mechanical energy into electrical energy.

dynamometer A machine which absorbs and measures the output power of an engine, usually on a test bed or rig.

59

E

earth fault indicator A set of lamps or an instrument calibrated in kilohms which is fitted on the main switchboard.

earthed (1) A fault in an electric circuit where a conductor is able to come into contact with the ship's hull or a metal enclosure. (2) A direct connection to the ship's hull which is required for metal frames or enclosures of electrical equipment.

earthed distribution system One pole or the neutral is connected to earth, i.e. the ship's hull. This system is not normally used on ships as an earth fault can result in the loss of essential equipment such as the steering gear.

easing gear A manually operated lever system for lifting a boiler *safety valve* from its seat.

eccentric A disc which is located on a shaft such that its centre and that of the shaft are offset. It acts as a kind of *cam* to convert a rotary motion into a linear motion.

echo sounder An instrument which measures the depth of water beneath a ship. Use is made of the time interval for ultrasonic waves to be reflected from the sea bed in order to determine depth.

economical speed The ship speed which will give the greatest annual return on the capital investment. It will be determined after a detailed economic analysis.

economizer A feedwater heater located within a boiler which uses heat energy from the exhaust gases.

eddy current A current which exists due to a voltage induced in a conducting mass by a varying magnetic field.

eddy-making resistance A componenet of the *residuary resistance* of a ship. It is due to the formation of eddies as water leaves the hull. See *frictional resistance, residuary resistance, wavemaking resistance*.

edge preparation The shaping of a plate edge for welding. This may take the form of a *V, X, Y* or *K* edge profile.

eductor See *ejector*.

effective power The power required to overcome the resistance of the ship at some particular speed.

efficiency The ratio of output to input for a system or operation. It is usually expressed as a percentage.

ejector A form of pump with no moving parts. A high pressure liquid or vapour discharges from a nozzle as a high velocity jet and entrains any liquid or gas surrounding the nozzle. See Figure E.1.

elastic coupling See *flexible coupling*.

elastic limit The maximum stress that can be applied to a metal without plastic deformation taking place. It is also known as yield point.

elastomer A synthetic material with properties similar to *rubber*.

60

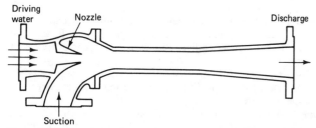

Figure E.1 Ejector

electric shock An effect produced on the human body by an electric current passing through it. The extent of the shock depends upon the value of the current, its duration, whether a.c. or d.c. and the path taken through the body.

electrochemical corrosion The wasting of metal as a result of an electrolytic cell being set up between two different metals and an electrolyte. The anode metal is eaten away. It is prevented by *cathodic protection.*

electrode (1) A conductor through which a current enters or leaves a liquid or gas. (2) The metal part of a welding rod used in electric *arc welding.*

electrolyte A liquid which permits the flow of current through it.

electrolytic corrosion See *electrochemical corrosion.*

electromagnet A ferromagnetic core of soft iron which is surrounded by a coil. A significant magnetic effect occurs only when a current flows in the coil.

electromagnetic flowmeter A flowmeter for liquids which conduct electricity. The liquid generates a voltage, as a result of it acting as a conductor in a magnetic field, which is related to its velocity.

electromagnetic induction The production of an electromotive force in an electric circuit by changing the magnetic flux which links with the circuit.

electromotive force See *voltage.*

electron A negatively charged elementary particle of an atom.

electronics A field of science and engineering which deals with electron devices and their use.

electrostatic An electrical phenomenon which is associated with electric charges which are independent of magnetic effects.

embargo A Government imposed ban on the movement of ships or cargo within a specific area.

embark To go on board a ship.

embedded temperature detector A *thermocouple, thermistor* or other small device which is fitted into an inaccessible winding or other part of an electrical machine to indicate operating temperature.

emergency bilge suction See *direct bilge suction.*

emergency fire pump A sea water pump which will supply the ship's *fire main* when the machinery space pump is not available. It will operate independently of all main power sources.

emergency generator A diesel driven generator of sufficient capacity to provide essential circuits such as steering, navigation lights and communications. It is an independent unit with at least two means of starting.

emergency procedure manual A book which details the action required if an emergency occurs on or near an offshore installation. Procedures for a variety of emergencies are included.

emergency supply A standby electricity supply in the event of main generating system failure. It may be provided from batteries or, more usually, by an *emergency generator* which feeds an emergency switchboard.

emulsion A suspension of one liquid within another such that they will not settle out. It is usually a reference to water in lubricating oil.

enclosed space entry The entry of any region where there may not be oxygen. A standard procedure must be adopted to ensure that the compartment can be safely entered.

enclosure A surrounding case or housing for electrical equipment which may be classified as *drip-proof, explosion proof, watertight, hoseproof* or *flameproof.*

end float The amount of permitted axial movement of a shaft.

end thrust The axial force on a *reaction steam turbine* rotor. It must be balanced out by a *dummy piston* or the use of double flow.

endoscope See *borescope.*

endurance limit The alternating stress which, during fatigue testing of a metal, will just produce fracture after a particular number of reversals.

energy spectra See *wave spectra.*

engine A machine for converting heat energy into useful mechanical work. It may be further described with regard to its operating cycle, e.g. *two-stroke* or *four-stroke,* or the source of energy, e.g. diesel or steam.

engine casing See *casing.*

engine indicator An instrument for measuring the power developed in the cylinder of an engine. A small piston moves in a cylinder against a calibrated spring. A drum oscillates in time with the engine piston movement and a *pv diagram* is drawn to a known scale. A calculation is required to determine the cylinder power developed. See *draw card.* See Figure E.2.

engine room log The record of main and auxiliary machinery parameters, tank levels, etc. A book may be used or a print-out from a computer.

engine trials (1) The various tests undertaken on main propulsion and auxiliary machinery during *sea trials.* (2) Testing of main propulsion machinery after overhaul with the ship moored.

engineer An officer who is qualified by training and examination to design, operate and maintain machinery. A marine engineer will usually hold a *Certificate of Competency* at some particular level, e.g. Class 1, Class 2.

Engineering Council A body which was established by Royal Charter in 1981 to advance education in, and to promote the science and practice of,

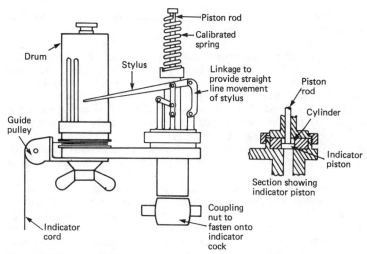

Figure E.2 Engine indicator

engineering (including relevant technology) for the public benefit and thereby to promote industry and commerce in the United Kingdom. The Engineering Council has authority to establish and maintain a register for the purpose of registering as professional engineers, as incorporated engineers or as engineering technicians such individuals as satisfy the Council of having achieved specified standards of education, training and experience. This function is carried out under the Engineering Council's authority by the Board for Engineers' Registration. The Council is located at 10 Maltravers Street, London WC2R 3ER, UK.

engineering economics The use of various evaluation techniques to produce a product design which has the highest standards of technical and economic performance.

Engineering Technician The following functional definition is used by the *Institute of Marine Engineers*:

The Engineering Technician (Marine) is competent by virtue of his/her education, training and practical experience to apply in a responsible manner proven techniques and procedures, and to carry a measure of technical responsibility, under the guidance of an *Incorporated Engineer* or a *Chartered Engineer*. He/she requires the ability to communicate clearly, both orally and in writing.

The Engineering Technician (Marine) must possess a recognized academic qualification, exemplified by an Ordinary National Certificate or City and Guilds Part II Certificate or equivalent TEC qualification in appropriate subjects, and have had a minimum of three years' engineering or shipbuilding experience of which two years must have been devoted to practical training.

entablature The structural element of a diesel engine in which the cylinders are located and onto which the cylinder heads are fitted. On a two-stroke slow speed diesel engine it is supported by the A-frames. See Figure S.11.

enthalpy A thermodynamic property of a fluid which is equal to the internal energy plus the product of pressure and volume. It is important in the study of flow processes where the heat taken in or rejected is equal to the work done plus the change in enthalpy.

entrance The immersed body of the ship forward of the *parallel middle body*.

entrepot An intermediate warehouse or port which is used for goods in transit.

entropy An aspect of a substance which will increase or decrease as the substance receives or rejects heat. Alternatively, it may be considered as a measure of the non-availability of the thermal energy of a system for conversion into mechanical work. It is often used as the horizontal axis of a graph where the vertical axis is absolute temperature and any area represents heat transfer, i.e. a *T–s* diagram.

epicyclic gears A system of gears where one or more wheels travel around the outside or inside of another wheel whose axis is fixed. The different arrangements are known as *planetary gear, solar gear* and *star gear*. Modern steam turbine gearing may be double or triple reduction and will be a combination from input to output of star and planetary modes. See Figure E.3.

epoxy resins Liquids which can be poured and cured at room temperatures. The cured material is tough, solid, durable and unaffected by oils and sea water. It may be used as a chocking material for engines, for adhesives or as a surface coating e.g. paint, because of its good adhesion.

equilibrium (1) The condition of a body which is at rest or moving with a uniform velocity. (2) A body upon which the resultant of all forces acting is zero.

equity crude The proportion of oil production which is retained by the company in a production sharing agreement between a state and a company.

erasable programmable read only memory A non-volatile computer memory which can be programmed and erased.

erection The process of positioning and welding units or parts of a ship's structure together on a berth or in a building dock.

ergonomics An aspect of *work study* which examines work in relation to the environment in which it occurs. The aim is to improve efficiency and the health and comfort of those working.

erosion A mechanical wastage process which removes metal and also enables corrosion to occur.

error The difference between an actual and the ideal or desired value or condition. See *deviation*.

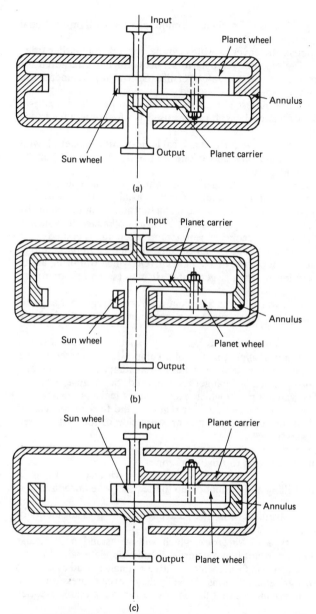

Figure E.3 Epicyclic gearing: (a) planetary gear, (b) solar gear, (c) star gear

65

error-rate damping The use of derivative action together with proportional action in a controller.

escape hatch A *hatch*, the cover of which can be opened from either side.

essential services The electricity supply to services which are required for navigation, propulsion and the safety of life. Special arrangements are made to ensure the supply is maintained.

ethane A colourless, odourless gas which is present in petroleum and is a major constituent of natural gas. It may be used as a fuel or as a feedstock source of ethylene.

ethylene propylene rubber An electrical cable insulation material with better resistance to moisture and ozone than butyl rubber. It can be damaged by oils and greases.

Euler's formula A mathematical relationship which gives the collapsing or critical load for a long straight column which is axially compressed.

evaporator (1) A device used to produce distilled water from sea water by either a boiling or a flash process. (2) A *heat exchanger* in which a refrigerant boils and the secondary refrigerant, often air, is cooled.

excise The duty charged on goods produced and sold in their home country.

exciter The source of all or part of the field current which will bring about electricity generation by an a.c. generator. It may be a static or rotary device.

exhaust gas The gas which leaves a system after an energy exchange or conversion process has taken place, e.g. from an engine cylinder or a boiler furnace.

exhaust gas boiler A boiler which is heated by exhaust gas from a diesel engine. On most motor ships it is part of a composite boiler, see Figure C.9. On diesel powered oil tankers, where an auxiliary boiler is already fitted, an exhaust gas heat exchanger may be used. See Figure E.4.

exhaust valve The valve through which exhaust gases leave an engine.

expansion valve A valve which meters or regulates the flow of refrigerant gas from the high pressure to the low pressure side of the system. See *thermostatic expansion valve.*

exploitation The entire process of oil production from survey through exploration and production to final marketing.

exploration A phase of *exploitation* which begins with searching in a particular area, drilling and evaluation up to the point when production is begun.

explosimeter An instrument for the detection and measurement of flammable gases in an atmosphere.

explosion proof A type of *enclosure* which will withstand a particular explosion within it and not ignite external flammable material.

explosion relief valve A *relief valve* which is fitted to engine crankcase doors as a safeguard against explosions. It will relieve excess pressure, stop flames being emitted and prevent the entry of air into the crankcase. See Figure E.5.

exposed location single buoy mooring A large *single point buoy mooring*. It is

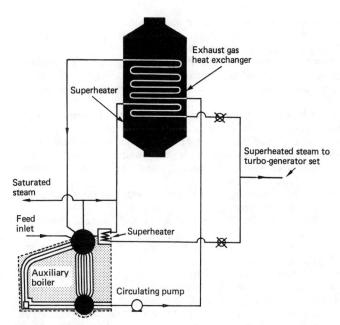

Figure E.4 Auxiliary steam plant system

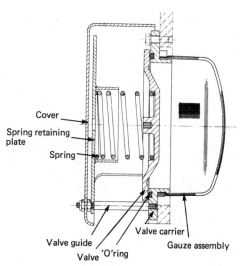

Figure E.5 Explosion relief valve

unmanned but has a helicopter landing deck fitted.

extensometer A measuring instrument which accurately determines the length of test pieces used in a *tensile test*.

Extra First Class Certificate The highest level of *Certificate of Competency* for an engineer. It is considered equivalent to a first degree and can be attempted by the holder of a Class 1 Certificate without further sea service.

Extra Master Certificate The highest level of *Certificate of Competency* for deck officers. It is considered equal to a first degree and can be attempted by the holder of a Class 1 Certificate without further sea service.

extraction pump This pump is used to draw water from a condenser which is under vacuum. It also provides the pressure to deliver the feed water to the deaerator or feed pump suction.

extreme breadth The maximum breadth over the extreme points port and starboard of a ship. See Figure P.7.

extreme depth The depth of a ship from the upper deck to the underside of the keel. See Figure P.7.

extreme draught The distance from the waterline to the underside of the keel. See Figure P.7.

extreme pressure lubricant A solid lubricant, e.g. graphite, or a liquid with additives which form oxide or sulphide coatings on exposed metal surfaces where, under heavy loading, the liquid film has broken down.

extrusion The shaping of metal, usually into a rod or tube cross-section, by forcing a block of the metal through appropriately shaped dies. Most metals are usually heated to reduce the extruding pressure required.

F

fabrication   The various processes which lead to the manufacture of structural parts for a ship.

factor of safety   The ratio of *ultimate* tensile stress to working stress for a material. It is always a value greater than one.

factor of subdivision   A value used in the calculation of permissible floodable length of a compartment with respect to damage stability of a ship. The value is determined by a formula which depends upon the length of the ship and a *Criterion of Service Numeral*.

Fahrenheit scale   A temperature scale in which the freezing point of water is 32 degrees and the boiling point is 212 degrees. Nine degrees on this scale is equal to five degrees on the Celsius scale.

fail safe   In the event of failure of a component or the loss of a supply or operating medium, the system or a device will take up a safe position or condition.

fair   To smoothly align the adjoining parts of a ship's structure or its design lines.

fairlead   An item of mooring equipment which is used to maintain or change the direction of a rope or wire in order to provide a straight lead to a winch drum. See Figure B.3.

falls   Wires or ropes which are used to hoist or lower a boat or cargo from a hold.

fan   A machine used to create mechanical ventilation. Various types are in use, e.g. centrifugal, axial flow and propeller.

farad   The unit of *capacitance* in the SI system of units.

fashion plate   A metal plate, fitted at the end of a superstructure deck and shaped with a curved edge, in order to minimize discontinuities and improve the appearance of the vessel.

fathom   A length of six feet, which is used to measure water depth.

fatigue   A type of failure occuring in metal as a result of a repeatedly applied fluctuating stress which may often be a lower value than the tensile strength of the material.

fault   A fracture in a rock formation where the two sides have been displaced with respect to one another and parallel to the fracture.

feathering   (1) The positioning of the blades of a controllable pitch propeller such that no thrust is produced. (2) The release of small quantities of steam by a safety valve at a pressure below the blow-off value.

feed check valve   A boiler feed water supply valve which is non-return and can also be regulated.

feed heater   A *heat exchanger* which increases the temperature of boiler feed water, usually by using some form of waste or exhaust steam which is condensed.

69

feed pump A high pressure pump which forces feed water into a boiler. It is often driven by steam from the boiler it supplies.

feed regulator A feed water control device which operates on the variation in water level alone. It is only suitable for small boilers operating at low pressure with small load changes.

feed tank A tank which stores a surplus or reserve of feed water in a boiler feed system.

feed water Distilled water which is supplied to a boiler and recirculates through the feed system. It is usually treated to remove air and impurities.

feedback A signal transmitted or fed back from a later to an earlier stage.

feeder (1) A vertical trunk which is filled with grain and temporarily fitted into the hatch of a hold containing grain. It will feed in grain as the hold contents settle. (2) The electrical lines which transmit power in a distribution system from switchboard to switchboard.

feedforward A supplementary signal transmitted or fed along a separate path, parallel to the main forward path, from an initial to a later stage.

feedstock The material which is fed into a plant in order to be processed.

feeler gauge A thin metal strip of a particular thickness, the value being marked upon it. It is used for the precise measurement of gaps.

fender A resilient device or material which is used to minimize impact or chafing damage to the side of a ship or other floating structure.

ferromagnetic A material, such as iron or cobalt, which has a high magnetic *permeability* and retains its magnetism in the absence of an external magnetic field.

ferrule A metal ring or cap which strengthens the end of a tube. It may be used as part of a joint or fixing arrangement.

ferry A *passenger ship* which operates on a regular scheduled service between ports. It may operate in a river, along a coast or as an ocean-going vessel.

fibre optics The use of very fine, optically insulated glass fibres to transmit light. The light may be used as part of a visual inspection system or as a control transmission medium.

field regulator A *potentiometer* which is used to adjust the voltage across the field of a d.c. *generator.*

filament lamp An electric light in which a filament is raised to incandescence by a current flow. The filament is housed in a glass bulb which is filled with an inert gas.

fillet A rounded corner created at an inside angle of a structure or casting.

filter (1) A device which mechanically separates solid contaminants from liquids. A filtering medium such as paper, wool or felt is usually used and must be cleaned or replaced when dirty. (2) A transducer which separates waves according to their frequency.

filter-separator A general duty air filter in a pneumatic supply system. It will remove the majority of water and any solid particles from the air passing through it. See Figure F.1.

final controlling element The element in a *control system,* whose action

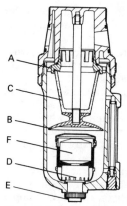

A Louvres
B Baffle
C Filter element
D Automatic drain assembly
E Waste pipe connector
F Float

Figure F.1 Filter-separator unit

occurs directly on the controlled body, process or machine, e.g. a valve.

fineness An indication of a ship's form in relation to the surrounding block. Low values of the various *form coefficients* indicate a fine ship.

finger plate A metal plate fitted to a machine to indicate the direction of rotation of the shaft or rotor.

finite element A small part of a large continuous structure which is being investigated with regard to its loading. Displacements, stresses and strains can be determined at the nodes where the various elements meet and the complete structure is thus analysed.

fire brick A *refractory* material which is used to line various parts of the furnace of a boiler. It is usually composed of alumina, silica and quartz.

fire detector A device which will detect some aspect of a fire, e.g. smoke, heat or infra-red light, and provide a signal to an alarm circuit.

fire main A sea water pipeline of sufficient diameter to provide an adequate supply for the simultaneous operation of two fire hoses. A fire main will exist in the machinery space and also on deck.

fire zone See *main vertical fire zones.*

firetube boiler An auxiliary *boiler* in which the hot gases pass through tubes which are surrounded by water. Low pressure, wet steam is produced.

fishing The recovery of lost tools or drilling equipment, i.e. 'fish', from a well.

fitting out The process of completing a ship by the addition of machinery and equipment.

Flag of Convenience A national flag of a country which provides virtually tax-free operation and other benefits to shipowners for minimal registration fees.

flame detector A device which can detect the ultra-violet or infra-red rays given off by a flame. It is used as a fire detector near to fuel handling equipment and at boiler fronts.

71

flame monitor A device which will detect the absence of a flame in a boiler furnace. It may also determine the quality of the flame and any detachment from the atomizer nozzle.

flame planer A machine which uses one or more oxy-acetylene burner heads to split or cut metal plates to a desired length or width by straight line cuts. *Edge preparation* for welding is also possible.

flame trap A gauze or perforated metal cover over an opening or vent to prevent the passage of flame.

flameproof A type of *enclosure* for electrical equipment which must withstand an explosion of any gas or vapour that enters. No damage must occur and no flammable matter must escape. This type of enclosure is used in *dangerous spaces*.

flammable A description of something that can readily be set on fire.

flammable limits The range of percentages of hydrocarbon vapour which can be ignited. It is between the lower flammable limit, 1%, and the upper flammable limit, 10%.

flange (1) The portion of a plate or bracket bent at a right-angles to the remainder. (2) To bend over at a right-angle. (3) A circular metal plate with holes formed or fitted on the ends of pipes in order to couple them together.

flap rudder A *rudder* with a separately movable flap at the trailing edge. It is operated at low or moderate ship speed to provide improved manoeuvrability.

flare (1) An outward curvature of the side shell at the forward end above the waterline. (2) A device which disposes of unwanted oil or gas by burning.

flash point The temperature at which a liquid heated in a particular type of apparatus will give off sufficient vapour to ignite momentarily upon application of a flame. A closed or open flash point value may be given to indicate the safe storage temperature of a fuel.

flashover An electrical discharge across the surface of an insulator.

flat A minor section of internal deck, often without sheer or camber. It may also be known as a platform.

flat compounded A d.c. *generator* with field windings arranged such that there is little or no variation in voltage between no load and full load.

flat margin A *double bottom* construction where the tank top extends horizontally to the ship's side.

flat of keel The width of the horizontal portion of the bottom shell, measured transversely. It may be called the flat of bottom. See Figure P.7.

flat plate keel The middle or centreline strake of plating in the bottom shell. It is increased in thickness for strength purposes and also to provide a corrosion allowance because it is difficult to paint this portion when the ship rests on docking blocks in a drydock.

Flettner rudder A special design of *flap rudder* which uses two narrow flaps at the trailing edge, one above the other.

flexible coupling A coupling or joining arrangement between rotating shafts which permits flexibility during start-up, some misalignment and also axial movement. Use may be made of rubber bushing, steel springs or membranes to transmit the load.

flexural vibration The vibration due to bending of a loaded member. It is of considerable importance when considering a ship and its cargo as a loaded beam supported upon waves.

flip-flop A circuit or device which contains active elements and may assume one of two stable states depending upon the nature of the input signal or which terminal last received the signal.

float-out The transportation of a completed offshore platform from the construction site to the drilling site.

floating battery A *battery* which is being charged and is also supplying current, i.e. discharging. A balanced situation is maintained with the battery remaining fully charged and available for peak loads.

floating dock A structure which can be ballasted to sink and receive a floating ship and then deballasted to bring the ship out of the water. It then functions as a drydock for ship maintenance while remaining *afloat*.

floodable length The maximum length of a particular portion of a ship which could be flooded without the *margin line* being submerged. See *factor of subdivision*.

flooding The entry of water into a tank or compartment either as a result of damage or in order to ballast the ship.

floor A transverse vertical plate fitted in a *double bottom*.

flotation barge A barge which is designed to carry a platform jacket to its offshore location. The jacket is placed vertically on the sea bed by controlled ballasting and sinking of the barge, which is then recovered.

flotsam Floating items which have been thrown overboard or lost in a storm or as a result of shipwreck.

flow line A pipeline of small diameter which feeds oil from one or more wells to a gathering centre.

flow nozzle A smooth bell-mouthed convergent entry fitting with a short cylindrical throat which projects downstream. Corner tappings are used to obtain differential pressure measurements in order to measure flow. This device is considered superior to the *orifice plate*.

flowchart A pictorial representation of the sequence and nature of operations that occur in a control system or computer program.

fluid coupling A device which uses the circulation of oil as the coupling arrangement. See *flexible coupling*. See Figure F.2.

fluidics The use of the interaction of flows of a medium to bring about control actions.

fluidized bed combustion The passage of a gas through a bed of solid particles to fluidize or cause the particles to float. Solid or liquid fuels may be burnt in such a bed to bring about highly efficient heat transfer.

flume A *roll stabilization* system using an athwartships tunnel or tank connecting two wing tanks containing a quantity of liquid. The liquid movement is out of phase with the roll motion and creates a moment

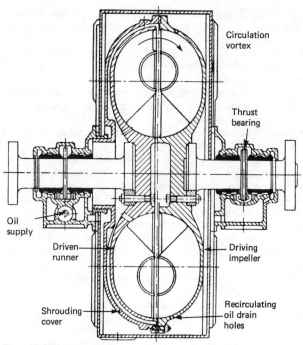

Figure F.2 Fluid coupling

which counteracts the roll.

fluorescence   The emission of light from a molecule that has absorbed light. During the interval between absorption and emission, energy is lost and light of a longer wavelength is emitted. This phenomenon is used in oil content monitors because oil fluoresces more readily than water.

fluorescent dye   A material which when added to a *heat exchanger* shell side liquid can be used to detect leaks in the tube plate. The leaking fluorescent liquid appears bright green when viewed with an ultra-violet lamp.

flush deck   A type of ship or the upper deck where it extends continuously from fore to aft. There is no *poop* but a raised *forecastle* may be fitted.

flying bridge   A navigating position, complete with magnetic compass, on top of the *bridge* or wheelhouse.

flywheel   A large disc or wheel which is fitted to the crankshaft of an internal combustion engine. It is usually made of cast iron or cast steel and acts as a store of kinetic energy to reduce speed variations when the engine is running.

74

foam A fire extinguishing medium which creates a mass of sticky bubbles to blanket a fire and exclude oxygen. It may be produced chemically or mechanically.

foaming (1) In a lubricating oil, this is the production of a froth of bubbles in the oil which will prevent correct lubrication. (2) In a boiler, this is the production of a froth in the steam drum at the water surface due to the presence of suspended matter, an excess of salts or oil. This will lead to priming, i.e. water carryover into the superheater.

following sea A sea which is moving along the same course as the ship.

foot valve A *valve* which is fitted at the lower end of a suction pipe. It is normally arranged to prevent a reversal of flow and a *strainer* may also be fitted.

footprint (1) The underwater area over which a submersible vehicle can operate. (2) The area occupied by an item.

force-balance The balancing of forces, usually in a pneumatic device, to achieve an equilibrium condition.

fore and aft Along the length of the ship.

forecastle The raised area of deck at the bow or the space between the raised deck and the deck below.

forefoot The curved part of the *stem* where it meets the keel.

Foreign Agreement Articles of agreement signed by the crew of a ship which is to trade Foreign Going as opposed to Home Trade.

foremast The forward mast where a ship has two or more masts.

forepeak A watertight compartment between the foremost watertight bulkhead and the *stem.*

forging The shaping of metal when it is hot but not molten.

form coefficient A value which describes the shape of a ship's hull in relation to a surrounding shape or block, e.g. *block coefficient.*

formation A unit of rock strata, usually one which has been penetrated during drilling.

formation testing Any means of determining the nature and quality of a reservoir, e.g. *drill stem testing.*

FORTRAN FORmula TRANslation. A high-level computer language used by scientists and engineers.

forward In the direction of, at, or near, the *stem.*

forward perpendicular An imaginary line drawn perpendicular to the waterline at the point where the forward edge of the *stem* intersects the summer *load line.*

forward shoulder The part of a ship where the *entrance* region meets the *parallel middle body.*

Fottinger coupling A type of hydraulic coupling. See *fluid coupling.*

fouling The covering of a ship's underwater surface with marine organisms such as green slime, weeds and barnacles.

founder To sink as a result of filling with water.

four-ram steering gear A hydraulically operated steering gear in which two diagonally opposite rams act to move the tiller. This arrangement provides a greater torque than a two-ram unit and the flexibility of different arrangements in the event of component failure.

four-stroke cycle An operating cycle for an internal combustion engine which requires four strokes or two revolutions of the crankshaft. See Figure F.3.

fraction A particular product obtained by distilling *crude oil* at a precise temperature and pressure. It is also called a cut.

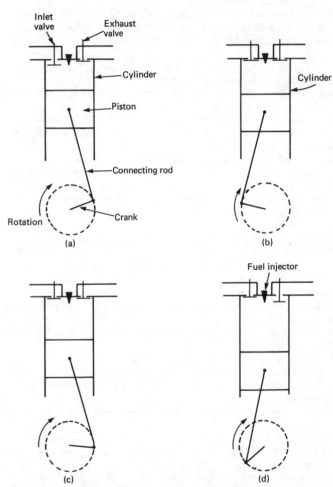

Figure F.3 Four-stroke cycle: (a) suction stroke, (b) compression stroke, (c) power stroke, (d) exhaust stroke

fractionation A distillation process in which different fractions are obtained. It is also called fractional distillation.

fracture A partial or complete break in a material. It may be further described by the nature of the surface at the break, e.g. brittle fracture.

frame A transverse structural member which acts as a stiffener to the shell and bottom plating.

frame bender A hydraulically powered machine which is used to bend the frames for a ship.

framing systems The various methods used to stiffen the bottom shell and side plating of a ship against the compressive forces of the sea. Two different types of framing are in general use or they may be combined. These are transverse, longitudinal and combined framing. See Figure F.4.

free surface A liquid surface in a partially filled tank which is free to move as the ship moves. The stability of a ship with free surface is reduced, regardless of the quantity of liquid in the tank.

freeboard The vertical distance from the summer load waterline to the top of the freeboard deck plating, measured at the ship's side amidships. See *Load Line Rules*.

freeboard deck The uppermost complete deck exposed to the weather and the sea. It must have permanent means of closure of all openings in it and below it.

Freedom vessel A popular ship design which served as a *Liberty ship* replacement.

freeing port An opening in a *bulwark* to enable water to flow off the deck into the sea.

freeing scuttle A flap fitted to some freeing ports which allows water to drain off the deck but not to enter.

freight A sum of money paid for the hire of a ship or for carrying goods by sea.

freight ton A cargo measure which may be 40 cubic feet or 2240 pounds. Freight is charged on this basis by using whichever provides the greatest payment.

freighter A general cargo ship.

freon A halogenated *hydrocarbon* usually given a number which is related to its chemical formula. It is used as a refrigerant.

frequency The time taken for a periodic disturbance to repeat. It is measured in hertz (cycles per second).

frequency modulation The variation of frequency of a *carrier wave* by the frequency of a modulating wave which is to be transmitted. The transmitted wave is extracted by demodulation.

frequency of encounter The frequency with which waves meet a ship which is moving through the water.

frequency response A measure of the ability of a device or system to respond to a cyclical input. A sine wave input is used and the system response to a change in frequency of the input is measured.

Fresh Water Allowance The fresh water *load line* or the increase in draught

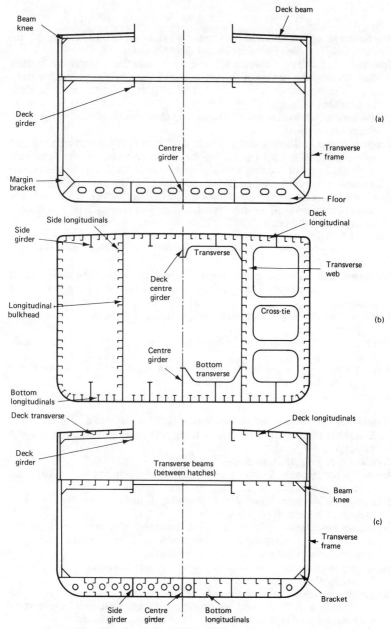

Figure F.4 Framing systems: (a) transverse, (b) longitudinal, (c) combined

permitted when a vessel operates in fresh water.

fresh water generator   An *evaporator* or *distiller*.

frictional resistance   The *resistance* experienced by a body moving through a fluid due to the velocity gradient across the boundary layer. For a ship it is considered to be dependent upon the wetted surface area, the length and type of the surface and the density of the liquid. See *eddy-making resistance, residuary resistance, wave-making resistance*.

Froude number   A dimensionless quantity, $V/\sqrt{(gl)}$, where V is the speed of the ship, l its length and g the acceleration due to gravity.

Froude's Law of Comparison   The wave or *residuary resistances* of geometrically similar ships are in the ratio of their displacements when their speeds are in the ratio of the square roots of their lengths.

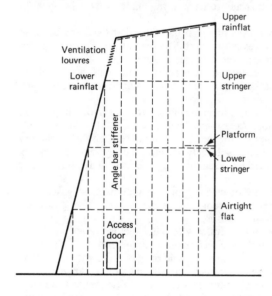

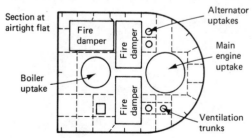

Figure F.5 A funnel

79

fuel blender A device which mixes heavy fuel and distillate in various proportions in the region of 70:30.

fuel coefficient A value used in conjunction with the *admiralty coefficient* formula in order to estimate fuel consumption for a vessel of known displacement at a particular speed.

fuel oil A product of the distillation of crude oil. Various grades exist from light to heavy. It is generally burnt in engines or boilers to utilize its heat energy.

fuel valve See *injector*.

full and down A description of a ship where the cargo space is full and the draught is equal to the appropriate *load line* mark.

full-bore safety valve A high capacity *safety valve* which is able to discharge four times as much steam as an ordinary spring loaded valve. See *high lift safety valve*.

full flow cargo system A means of cargo handling that has been used on tankers. Large sluice valves allow the oil to flow aft to the cargo pump suctions. No suction piping is used in this system.

full-wave rectifier A device which inverts alternate half waves of an alternating input so that the output contains two half wave pulses for each input cycle. See *rectifier*.

fullness An indication of a ship's form in relation to the surrounding block. High values of the various *form coefficients* indicate a full ship.

funnel A surround and support for the various uptakes from the machinery space of a ship. Its shape is largely determined by smoke clearing requirements and the need for streamlining to reduce air resistance. See Figure F.5.

funnel gas inerting The use of cleaned inert boiler exhaust gases as a blanket cover over the cargo tanks of an oil tanker. The gases are scrubbed, dried and filtered before use.

furnace The interior region of a *boiler* where the fuel is burned.

fuse A device which protects a circuit from overcurrent. It comprises a fusible length of wire enclosed in a ceramic or glass tube with metal end contacts. The rating is the continuous current that can be carried.

fused isolator An isolating switch that incorporates a *fuse*. When the switch is opened the fuse can be safely replaced. It may be interlocked with the cubicle door handle.

G

gag   A device which holds one safety valve shut while the other is being set.

gain   The ratio of output to input signals, which are of the same physical form, in any part of a control system.

gallon   A unit of volume which is the space occupied by 10 pounds of distilled water. A US gallon is related to 8.3359 pounds of distilled water.

galvanic corrosion   See *electrochemical corrosion.*

galvanizing   The coating of iron or steel with zinc, usually by dipping in a bath of hot zinc and flux. The iron or steel is thus protected against atmospheric corrosion.

galvanometer   An instrument used for measurement or indication of a small direct current flow.

gangway   A ramp or stairs arrangement which is used for embarking or disembarking.

gantry   (1) The platform or bridge structure of a travelling crane. (2) An enclosed gangway for embarking or disembarking passengers.

gap press   A hydraulically powered press which cold works steel plate. It will bend, straighten, dish or swedge by the use of different die blocks. It is also known as a *ring press.*

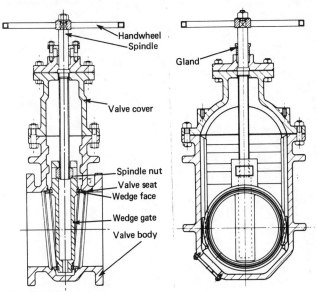

Figure G.1 Gate valve

81

garboard strake The bottom *shell plating* on either side of the *keel* plate.

gas cap The volume of gas at the highest point in a reservoir rock. See Figure A.3.

gas freeing The removal of dangerous gases from a cargo tank after discharging the cargo.

gas lift The use of injected gas to aerate the oil in a production well and thereby increase the flow.

gas oil A distillate of petroleum which is used as a fuel in high speed diesel engines and heating installations. It is heavier than kerosine and lighter than light lubricating oil.

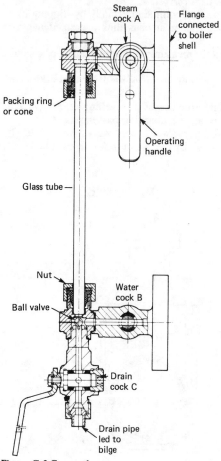

Figure G.2 Gauge glass

gas/oil ratio The volume of gas, measured at atmospheric pressure, which is produced per unit volume of oil produced from a well.

gasket A joint, usually of flexible material, which is positioned between metal surfaces to prevent leakage.

gasoline A distillate of petroleum which is used in spark ignition internal combustion engines. It is also known as *petrol*.

gate A logic element which operates on an applied binary signal. The gate will have a specified logical function, e.g. AND, OR, NOT.

gate valve A valve which permits uninterrupted flow when open. The gate can be raised clear of the valve opening and may be parallel or wedge-shaped in section. See Figure G.1.

gauge glass A water level gauge which is fitted to a boiler. Low pressure auxiliary boilers use a tubular gauge glass. High pressure watertube boilers use a *plate type gauge glass*. See Figure G.2.

gauss A unit of magnetic flux density. The SI unit is the tesla which is equal to 10 gauss.

gear pump A displacement *pump* comprised of two interconnecting gear wheels in a closely fitting casing. The liquid is trapped between the gear teeth and the casing as one gear wheel is driven and the other meshes with it.

gear train A system of gear wheels which transmit motion from one shaft to another. Various arrangements exist, e.g. single reduction, double reduction.

gear wheel The wheel or hub on which the gear teeth are cut. The teeth may be parallel to the axis of rotation (*spur gear*) or inclined at an angle (*helical gear*).

gearbox The casing which encloses a gear train or the complete assembly.

general average A principle relating to an action, sacrifice or expenditure which is taken or made for a common good or safety, e.g. the jettisoning of cargo from a ship which is aground in order to refloat it. It relates to damage incurred by a ship or its cargo or expenses related to a maritime adventure. The loss or expense is shared (adjusted) amongst those who benefitted. See *average adjuster.*

general cargo ship A vessel designed and built for the carriage of general cargo. A number of large clear holds will exist and one or more separate or tween decks may be present within the holds. The hold openings are secured by hatch covers. See Figure G.3.

General Council of British Shipping The central organization which represents UK shipowners in all matters related to seagoing personnel.

general lien A right to retain goods against settlement of a charge which is extended from a particular cargo to cargo carried on other vessels.

generator A machine which converts mechanical power into electrical power. It is usually further described by the type of electricity produced, i.e. a.c. or d.c. generator.

gill jet thruster A *thruster* device which uses a vertical axis propeller in a T-shaped tunnel. Water is drawn in from both sides and leaves through

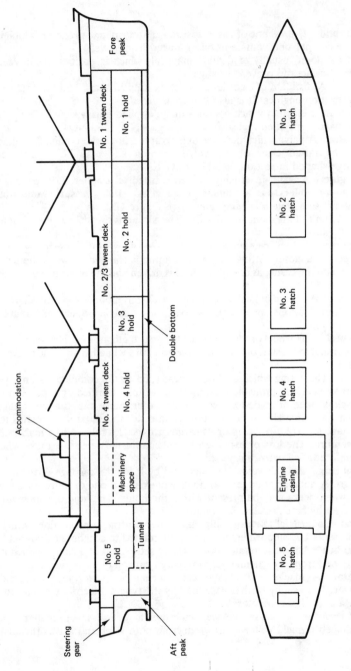

Figure G.3 General cargo ship

the bottom of the hull. Rotatable gill fins direct the water in one of a number of fixed positions around a circle.

girder   A continuous stiffening member which runs fore and aft in a ship to support the deck.

gland sealing   The use of mechanical or other means to prevent leakage around a shaft or rotor, e.g. at the ends of a steam turbine rotor. See *mechanical seal, stuffing box.*

gland steam condenser   A shell and tube type *heat exchanger* which is used in a *closed feed system* to collect and condense steam and vapour from the turbine gland steam system.

glass reinforced plastic   A combination of thin fibres of glass in various forms which, when mixed with a *resin*, will cure (set) to produce a hard material which is strong and chemically inert. It has a variety of uses for general repairs.

globe valve   A *valve* with a spherically shaped body enclosing the valve seat or disc. Liquid flow is arranged from below to above the valve seat so that the upper chamber is not pressurized when the valve is closed.

go-devil   (1) A device which is dropped down a well in order to start the operation of a tool or detonate an explosive charge or device. (2) Another name for a *pig*.

goose neck   (1) A fitting on the end of a boom or derrick which connects it to the mast or post and permits a swivel motion. (2) A bend in the end of an air pipe or vent such that the open end points downward.

gouging   The removal of metal from a welded seam in order to make a back-run for a butt weld. An arc-air arrangement or a modified gas cutting torch may be used.

governor   A controller which is used to maintain the rotational speed of an engine within certain prescribed limits throughout the operating power range. In addition to speed control, modern governors also provide load control and fuel limitation to protect an engine under adverse operating conditions.

grab   A bulk cargo handling device consisting of a bucket which is hinged to open and close. It is open when lowered into the cargo and then closed around the material and lifted out full. It is opened to discharge the cargo ashore.

graduation   The marking or setting out of a scale.

grating   A platform or part of a platform made of a grid arrangement of steel strips. It is often used in the machinery space.

gravimetric method   A type of geophysical survey which measures differences in the force of gravity between sub-surface rocks. A sensitive measuring instrument called a gravimeter is used.

graving dock   A *dry dock.*

gravity davits   *Davits* which slide down and position a lifeboat for lowering as soon as they are released. See Figure G.4.

gravity drainage   The flow of oil towards the borehole when the gas drive has ceased. Primary production is at an end and some form of secondary

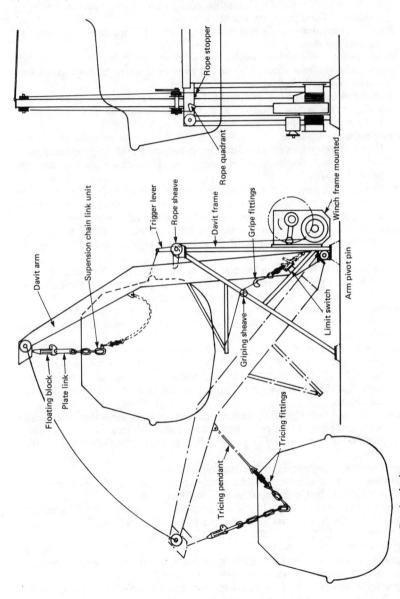

Rope stopper

Rope quadrant

Davit arm

Supension chain link unit

Trigger lever

Rope sheave

Davit frame

Gripe fittings

Winch frame mounted

Floating block

Plate link

Griping sheave

Limit switch

Arm pivot pin

Tricing fittings

Tricing pendant

Figure G. 4 Gravity davit

recovery is required.

gravity platform An offshore *platform* which may be used for storage or storage, development drilling and production. Oil or ballast is always present in the cellular base or stabilizing columns.

gravity welder A device consisting of a tripod, one leg of which acts as a rail for a sliding electrode holder. Once positioned and with the arc struck the device welds automatically until the holder operates a trip at the bottom of the rail.

green sea A sea which travels over the deck of a ship without any waves breaking. Considerable damage may be done to the ship by such a sea.

Greenwich Mean Time The mean solar time at the Greenwich meridian. It is the standard time for navigation and astronomical observations.

gripes Wire ropes which secure a lifeboat against the cradle when it is up on the davits. See Figure G.4.

grommet A ring of soft material positioned beneath a nut or bolt head to create a watertight joint.

gross tonnage The total of the underdeck *tonnage* (a measure of volume) and the tonnage of any tween deck spaces between the second and upper decks, any enclosed spaces above the upper deck, any excess of hatchway over 0.5% of the gross tonnage and optionally any engine light and air spaces on or above the upper deck. The International Convention on Tonnage Measurement of Ships 1969 requires the use of a formula related to the total volume of all enclosed spaces.

ground speed The speed of a ship relative to the land. It is measured on trials and must be corrected to find the speed through the water.

grounded Connected to earth, or an extended conducting body, e.g. the hull of a ship, which acts as an earth.

guardian valve An astern steam isolating valve. It is one of the *manoeuvring valves* for a steam turbine.

gudgeon A solid lug on the sternframe or rudder with a suitable opening to receive the *pintle*.

gudgeon pin The pin which secures the piston to the small end of the connecting rod. See Figure T.7.

guide and slipper An assembly fitted to a slow speed, two-stroke diesel engine to transfer the connecting rod side thrust to the engine framework. The slipper or crosshead shoe is part of the crosshead and piston rod assembly and slides in a guide in the *A-frame*.

guide lines Wires, extending from the drilling slot of a floating platform to a guide base on the sea bed, which guide equipment down to a sub-sea well-head.

guillotine A hydraulically powered shearing machine which cuts small items of plate such as brackets for a ship's structure.

gun perforater A device which fires bullets or explosive charges into casing in order to allow oil into the *borehole*.

gunmetal A type of *bronze* containing a proportion of zinc between 2 and 4%. Lead and nickel may also be added. It is used for cast fittings where resistance to corrosion is required, e.g. valve bodies.

gunwale The upper edge of a ship's side where the *sheer strake* meets the deck plating.

gusset plate A bracket plate usually positioned in a horizontal or almost horizontal plane.

gutter A channel on either side of the deck which collects and directs water into the *scuppers*.

guy A rope which holds or steadies a derrick or boom in position.

gypsy A small drum or warp end which may be fitted on a *winch* or *windlass*.

gyro compass A gyroscope which is electrically rotated and controlled such that its axis of rotation is always on the meridian.

H

Hague Rules International legislation adopted in 1972 which regulates the carriage of goods by sea.

half-breadth The distance across a half section of a ship.

half-breadth plan A drawing of a ship which shows the shapes of the waterlines and decks formed by horizontal planes at the various waterline heights above the keel. It is part of the *lines plan*. See Figure L.3.

half-wave rectifier A device which only permits alternate half waves of an alternating input to pass and thus the output contains one half wave pulse for each input signal. Each pulse is unidirectional.

Halon A halogenenated hydrocarbon gas which is used as a fire extinguishing agent. Two examples are Halon 1301 (BTM) and Halon 1211 (BCF).

hand steering gear An emergency steering arrangement located aft. Rods and gear wheels enable operation of the steering gear from an after deck position.

harbour A sheltered haven in which ships may anchor or moor. It may be a natural geographical formation or artificial.

harbour dues Charges which are levied upon a ship by the harbour authorities for the use of the harbour.

hard copy A permanent record of computer output, usually in a printed form.

hardening A heat treatment process for *steel* where it is heated to 850–950°C and then rapidly cooled by quenching in oil or water. The hardest possible condition for the steel is produced and the tensile strength is increased.

hardness (1) The ability of a material to resist plastic deformation, usually by indentation. (2) A measure of the ability of a sample of water to produce a lather with a soap solution. It is an indication of the presence of dissolved salts.

hardware The physical equipment that comprises a computer system.

harmonic A sinusoidal oscillation whose frequency is some multiple of the fundamental frequency.

hatch See *hatchway*.

hatch beam A removable beam which is fitted over a hatch opening, usually beneath a wooden hatch cover.

hatch coaming Vertical plating which surrounds a hatch opening. Its height is determined by the *Merchant Shipping (Load Line) Rules 1968*. The hatch cover rests, and is secured, on the top of the hatch coaming.

hatch cover A wooden, or more usually steel, cover which creates a watertight hatch to protect the cargo and also stiffens up the structure of the opening. See Figure H.1.

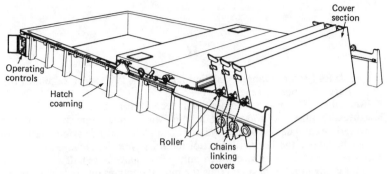

Figure H.1 Hatch cover

hatchway   The opening in a deck or deckhouse which provides access to the various tween deck and hold spaces below.

hawse block   A wooden plug which fits into the *hawse pipe* to stop sea water coming up and onto the deck.

hawse pipe   A thick section pipe through which the anchor cable passes from the forecastle deck to the ship's side. A doubling plate is fitted around it at the forecastle deck and a chafing ring at the ship's side.

hazard zone   A working area in which a high risk of fire or explosion exists. On oil rigs four types numbered 0, 1, 2 and 3 may exist with 0 indicating flammable vapours are always present and 3 a safe area.

head   The height of a column or body of water which represents a pressure acting at the datum point.

head loss–flow characteristic   A graph of head against flow for a pump on which is drawn the pump characteristic, the system characteristic and the net positive suction head available and required. See Figure H.2.

header   A solid drawn circular section tube in which the various waterwall tubes terminate in a *watertube boiler*.

header tank   A liquid storage tank which maintains a head or gravity pressure on the system. It acts as an expansion tank and will also supply liquid to make up for system losses.

heat   A form of energy which is in the process of transferring from a system to its surroundings as a result of temperature difference.

heat balance   A statement, usually in the form of a system diagram, of the energy available in a system and how it is all distributed. Details of pressure, temperature and mass flow of, for example, a steam turbine cycle would be given at the various points in the system.

heat detector   A device which will detect a fire, or a considerable change in temperature, as a result of the action of heat on a sensing element.

heat exchanger   A device in which cooling or heating takes place, usually without actual contact between the flowing substances. In construction may be a *shell and tube* or a *plate type heat exchanger*.

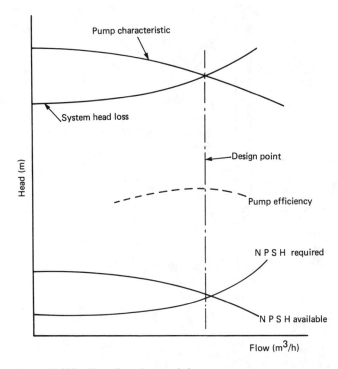

Figure H.2 Head loss–flow characteristic

heave compensator A mechanism which is used to protect the *drill string* and *blow out preventer* stack from the heaving motion of a floating drilling platform. It is usually hydraulic or hydro-pneumatic in operation. See Figure D.2.

heaving The up and down linear motion of a ship in the sea. See Figure R.6.

heavy crude Crude oil with a high proportion of heavy oil *fractions*.

heavy lift derrick A cargo handling device, usually of patented design, for large heavy items, e.g. *Stulken derrick*. See Figure H.3.

heavy lift vessel A vessel designed, constructed and outfitted in order to lift very large masses such as small ships or accommodation modules for oil platforms.

heel (1) To incline or *list* in a transverse direction due to an external force such as the wind or sea or an internal force such as shifting cargo. (2) The base of a *mast* or *derrick*. (3) The position where the *keel* and the base of the *stern frame* meet.

91

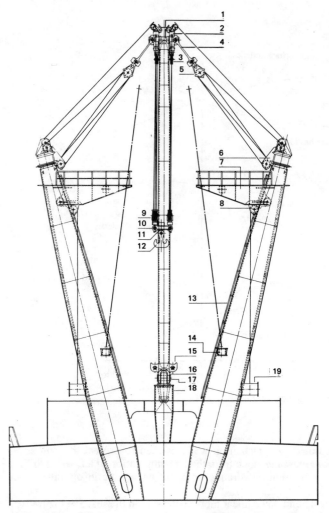

1 Derrick head fitting.
2 Pendulum block fitting with guide rollers.
3 Upper cargo blocks.
4 Connecting flats.
5 Lower span block.
6 Span swivel.
7 Cross-tree.
8 Inlet for the hauling part.
9 Lower cargo blocks.
10 Connecting traverse.
11 Swivel eye for flemish hook.
12 Flemish hook.
13 Ladder.
14 Gooseneck pin socket.
15 Fastening device for lower cargo block.
16 Heel fitting.
17 Derrick pin.
18 Gooseneck and gooseneck pin socket.
19 Winches.

Figure H.3 Heavy lift derrick

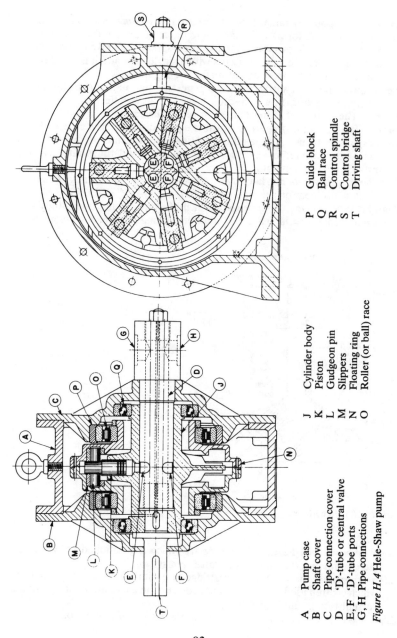

A	Pump case	P	Guide block
B	Shaft cover	Q	Ball race
C	Pipe connection cover	R	Control spindle
D	'D'-tube or central valve	S	Control bridge
E, F	'D'-tube ports	T	Driving shaft
G, H	Pipe connections		

J	Cylinder body
K	Piston
L	Gudgeon pin
M	Slippers
N	Floating ring
O	Roller (or ball) race

Figure H.4 Hele-Shaw pump

93

Hele-Shaw pump   A radial type of *variable stroke pump*. See Figure H.4.

helical gearing   A form of gearing where the teeth are cut on a helix to the rotational axis. In this way tooth loading is spread out and a quieter power transfer occurs. End thrust occurs with a single gear and it is usual to use double gears with opposite cut teeth. See Figure D.4.

helicoidal surface   A surface which is generated by rotation and translation in relation to an axis, e.g. a screw thread or a propeller.

henry   The SI unit of either mutual or self *inductance*.

hertz   The SI unit of *frequency* which is equal to one cycle per second.

high expansion foam   A type of fire fighting foam which is produced from a liquid concentrate. It expands up to one thousand times after passing through a foam generator.

high-level programming language   A computer programming language which is problem or procedure oriented, e.g. *BASIC* or *FORTRAN*.

high lift safety valve   A *safety valve* which lifts quickly and opens fully as a result of some special feature or additional lifting surface. See Figure H.5.

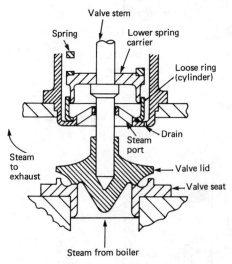

Figure H.5 High lift safety valve

high pressure feed heater   A *feed heater* which is positioned between the feed pump and the boiler in a *closed feed system.*

high rupturing capacity fuse   A cartridge *fuse* with a high breaking capacity which is often used for overcurrent protection in a circuit. Various normal current ratings are available.

high seas   The seas outside of the territorial waters of any nation.

94

high speed engine   A compression ignition engine with a rotational speed in excess of 700 rev/min.

high voltage   An electrical distribution system operating at either 3.3 kV or 6.6 kV. On oil platforms this may even be as high as 11 kV.

higher calorific value   See *calorific value.*

higher tensile steel   A *steel* with an increased tensile strength and adequate notch toughness, ductility and weldability. It has particular application in ship's structures where high stresses occur.

hogging   The condition of a floating ship when the distribution of weight and buoyancy along its length is such that the buoyancy amidships exceeds the weight.

hoist   To raise a load.

hold   The lowest cargo stowage compartment in a ship.

holding down bolts   The securing bolts, which are usually fitted, for main propulsion machinery and thrust blocks.

hole   See *borehole.*

home port   The port where a vessel is registered. The name is painted on the stern. See *Certificate of Registry.*

Home Trade   Trade which takes place within the British Isles and the European ports as far as Brest and the River Elbe.

horn   The part of a *stern frame* on which a spade type rudder is hung.

hoseproof   A form of *enclosure* for electrical equipment which indicates that water from a nozzle in any direction and under specified conditions will have no harmful effect.

hot spot   A region of metal which has considerably increased in temperature

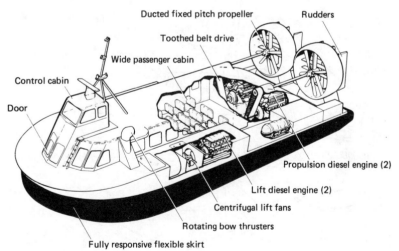

Figure H.6 Hovercraft

95

due to friction or a lubrication failure. It may bring about the ignition of a flammable vapour, e.g. lubricating oil vapour in a crankcase.

hotwell A feed water tank in an open feed system. It collects condensate from the condenser and the main feed pump draws from it.

hovercraft A ship which is supported by an air cushion trapped beneath it by a curtain or skirt. Propulsion may be by airscrews or propellers. See Figure H.6.

hull The structure of a ship comprising the shell plating, framing, decks, bulkheads, etc.

hull efficiency The ratio of *thrust power* to *effective power*, which is usually a value greater than one.

humidity A measure of the amount of water vapour in a given volume of gas. Most measurements relate to air. See also *relative humidity* and *absolute humidity*.

hundred year storm A criterion used in the design of offshore oil rigs and meant to represent conditions that will only occur once in a hundred years.

hunting The prolonged oscillation or cycling of a controlled variable about a specified reference value.

hunting gear The feedback mechanism of a *steering gear* which repositions the floating lever of the hydraulic pump as the tiller moves to the desired position.

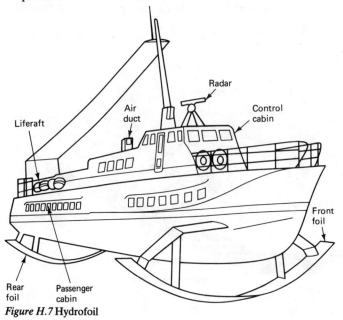

Figure H.7 Hydrofoil

96

hybrid platform A gravity structure, usually a *production platform* where steel and concrete are used in the construction in almost equal quantities. The tower, base and modules can usually be built separately and assembled on site.

hydrant The terminal point of a water main which has fittings for the attachment of a hose pipe.

hydraulic balance An arrangement within a *turbofeed pump* which controls the axial movement of the impeller assembly using some of the discharging liquid in a balance chamber.

hydraulic control The use of a flowing liquid as the control medium in pumps, motors, valves, actuators and ancillary fittings.

hydraulics The science dealing with the flow of liquids, usually in the transmission of power.

hydrazine A colourless alkali which is used in boiler feed water to remove oxygen.

hydrocarbon An organic compound of hydrogen and carbon which occurs in many forms in *crude oil* and *natural gas.*

hydrocracking A *catalytic cracking* process which takes place in the presence of hydrogen. Lighter distillates, in particular petrochemicals and aromatics, are produced.

hydrofoil (1) A *dynamically supported craft* which uses hydrofoils to create a lifting force such that the hull is clear of the water. It is a high speed craft which is usually used as a ferry. See Figure H.7. (2) An immersed metal plate of streamlined cross-section which creates a hydrodynamic lift when moved through the water.

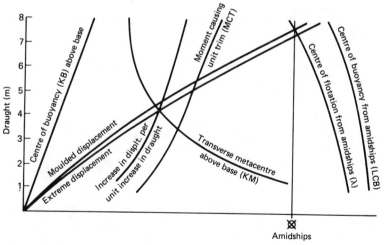

Figure H.8 Hydrostatic curves

97

hydrographer A person who surveys and produces charts of the oceans and seas for navigation purposes.

hydrometer A float which is suitably weighted and calibrated to measure the density of liquids, e.g. sea water, battery acid.

hydrophone An instrument which detects sound waves passing through water. It is used in *seismic surveying*.

hydrostatic curves A series of graphs drawn to a vertical scale of draught and a base of length which give values such as centre of buoyancy, displacement, moment causing unit trim and centre of flotation. See Figure H.8.

hydrostatic release A mechanism which will release at a pre-determined depth of water. It is used to secure *liferafts* and thus enable them to float free from a sinking ship.

hygrometer An instrument which measures the *humidity* of air or a gas.

hypalon A cable sheath material (chlorosulphonated polyethylene) which has a good resistance to mechanical damage, acids and alkalis and is flexible.

hyperbaric chamber A compartment which can be sealed and pressurized to reproduce conditions at various water depths. It may be used to enable divers to decompress or sometimes for dry work in deep water.

hysteresis The internal energy loss in an element that results in an output signal which depends not only on the input signal but whether it is increasing or decreasing in value.

I

ice breaker   A vessel which has been specially designed and strengthened in its construction so that it can break up ice which has formed in navigable channels.

Ice Class notation   A *classification* given to vessels which have additional strengthening to enable them to operate in ice bound regions.

Igema gauge   A remote reading boiler water level indicating instrument. It is a U-tube *manometer* containing two different coloured liquids.

ignition   The setting on fire of an explosive mixture, usually within an engine cylinder.

ignition delay   The time period between the beginning of fuel injection into an engine cylinder and the start of combustion.

ignition quality   See *diesel index number*.

immersion   The change in draught resulting from the addition or removal of a particular mass. In SI units it is called tonnes per centimetre immersion, TPC.

impedance   The ratio of a sinusoidal voltage and current where both are expressed in *root mean square* terms.

impeller   The rotating element within the casing of a *centrifugal pump*. Liquid enters at the centre or 'eye' and is discharged at the perimeter.

imports   All goods and commodities brought into a country which originate in other countries.

impulse turbine   A steam turbine whose blades are rotated by an impulsive force as a result of changing steam velocity.

in phase   A description of alternating quantities of the same *frequency* which reach corresponding values simultaneously.

in-water survey   The *survey,* usually of a very large crude carrier, which is undertaken in the water. Hull plating must be cleaned, normally by a diver operated machine, and then a remotely operated underwater TV camera is used to view the ship's plating.

inboard   In a direction towards the centreline of a ship.

incinerator   A waste disposal unit which will burn solid material and all types of liquid waste.

inclination   The angle of *list* of a ship, measured from the vertical.

inclined manometer   An instrument which is used to measure low pressures. The applied pressure acts on a container of liquid and forces some of it along an inclined glass tube.

inclining experiment   An experiment, conducted on an almost completed ship, in order to determine the vertical position of the *centre of gravity*. The ship is inclined or heeled by moving weight transversely across the deck.

inclinometer   See *clinometer*.

99

incombustible See *non-combustible*.

Incorporated Engineer The following definition would refer to an Incorporated Engineer (Marine):

The Incorporated Engineer (Marine) is competent by virtue of his/her education, training and subsequent experience to exercise independent technical judgment in and assume personal responsibility for duties in the marine engineering field.

His/her education and training is such that by the application of general principles and established techniques, he/she is able to understand the reasons for and the purposes of the operations for which he/she is responsible.

The Incorporated Engineer (Marine) performs technical duties of an established or novel character either independently or under the general direction of a *Chartered Engineer* or scientist. He/she requires the power of logical thought and when in a management role the quality of leadership. His/her work is at a higher level of responsibility than that of an *Engineering Technician*.

The Incorporated Engineer (Marine) has an academic qualification of a standard not lower than that exemplified by a Higher National Certificate or a City and Guilds Full Technological Certificate or equivalent TEC qualification in appropriate subjects and has had a minimum of five years' engineering or shipbuilding experience of which two years must have been devoted to practical training. The title Technician Engineer was previously used.

indemnity A security against, or compensation provided for, loss or damage.

index The indicator which, by its position in relation to a scale, provides a value of the measured quantity, e.g. pointer, recording pen or stylus.

indicated power The power developed in an engine cylinder. It is determined from an indicator diagram and certain basic information about the engine. See *engine indicator*.

indicating instrument A measuring instrument in which the value of the measured quantity is visually indicated, but not recorded.

indicator diagram The *pressure-volume diagram* produced by an *engine indicator*.

indirect drive The use of a *gearbox* between a high or medium speed main propulsion engine and the propeller shafting.

induced draught An air and gas flow through a boiler furnace which is created by a fan in the exhaust uptake.

inductance The property of an electric circuit whereby a varying current induces an electromagnetic force in the circuit or a nearby circuit.

induction motor An a.c. motor in which the current in the rotor is created by electromagnetic induction.

inert gas (1) Gaseous elements which do not chemically combine, e.g.

100

helium, neon and argon. (2) A mixture of gases which contain less than 5% oxygen and will not therefore support combustion.

inert gas generator A device, somewhat like a boiler, in which fuel is burnt to create exhaust gases which contain less than 5% oxygen. The inert gas is used for fire extinguishing.

inert gas plant A system which cleans exhaust gas from a boiler or engine and supplies it to a distribution main. The inert gas, which contains less than 5% oxygen, is used as a 'blanket' to cover oil cargoes during passage and discharge.

inflammable See *flammable* (preferred term).

information technology The various techniques associated with information handling. This may include the use of computers and telecommunications systems to handle, store or process data in pictorial, audio or numerical form.

infra-red radiation Invisible electromagnetic rays of wavelength 0.75 to 1000 micrometres. They are perceived as heat and may be utilized in temperature measurement.

inherent regulation A process property which results in equilibrium after a disturbance without any monitoring feedback.

initial stability The *stability* of a vessel in the upright condition which is determined by the *metacentric height,* GM. If the metacentre, M, is above the centre of gravity, G, the GM is said to be positive and the vessel is stable.

injection The process of spraying fuel into an engine cylinder by the use of an injector.

injector A device which receives pressurized fuel as a liquid and sprays it into an engine cylinder as a fine mist. It consists of a nozzle and a nozzle holder or body. The nozzle has a series of small holes around its tip through which the fuel is sprayed into the engine cylinder. The term *fuel valve* is also used. See Figure T.7.

Inmarsat International Maritime Satellite Organization. A communications network available to ships worldwide. It provides telephone, telex and distress message transmission facilities.

input Data or programs which are received by a computer from an external source.

input/output device A unit which is used to transmit data to a computer or receive information from it.

input signal The signal which, when received by an element, results in some action.

insert plate A steel plate of greater thickness which is fitted at a region of increased stress, e.g. a hatch corner.

Institute of Chartered Shipbrokers A professional body whose membership is open to persons and companies involved in shipbroking. It was founded in 1911 and is managed by a Controlling Council of Fellows. The headquarters are at 24 St Mary Axe, London EC3A 8DE, UK.

Institute of Marine Engineers A learned society founded in 1889 to promote the scientific development of all aspects of marine engineering. It is a Nominated and Authorized Body of the *Engineering Council* and members are entitled to be registered as *Chartered Engineers*. The headquarters are at 76 Mark Lane, London EC3R 7JN, UK.

Institute of Petroleum An organization, based in the United Kingdom, which is sponsored by the oil industry. It promotes the study of petroleum and its allied products and is the recognized standards authority for testing methods. The headquarters are at 61 New Cavendish St., London W1, UK.

instruction A set of characters which specify an operation to be performed as a unit of a *program*.

instrument A device which provides a measurement or indication of the quantity measured.

instrument air Compressed air which is suitable for use in pneumatic equipment. It must be free of oil and dust and dry enough to ensure that no water condenses anywhere in the system.

instrument range The range of values over which an *instrument* is able to measure.

instrument transformer A *transformer* which produces a current or voltage in the secondary that is maintained in a particular proportion, with respect to magnitude and phase, to the primary.

instrumentation amplifier A form of *operational amplifier* which has internal feedback. It is used to amplify a differential input signal and reject any noise and interference. It is usually an *integrated circuit*.

insulated a.c. system A *distribution* system which is electrically insulated from a ship's hull. This is the usual arrangement on board ship as an earth fault would not result in the loss of any essential equipment.

insulation (1) The confining, by any means, of a transmissible phenomenon to a particular path or location, e.g. heat, vibration, electricity. (2) Any material which is used to achieve (1).

insulation resistance The resistance of electrical insulation measured either between a conductor and earth or between conductors. It is usually measured using an instrument such as the *Megger tester*.

insulator Any item, made of an insulating material, which is used to provide insulation and, usually, some form of mechanical support.

integral action The action of a control element where the output signal changes at a rate proportional to its input signal. It is also called *reset action*.

integral action time In a proportional plus integral controller, this is the time interval in which integral action increases by an amount equal to the proportional action signal, where the deviation is constant.

integrated circuit An electronic circuit formed on the surface of a tiny silicon chip.

integrated monitoring system The combining of all the individual control systems in a plant into a single computer controlled system. This would include all aspects of navigation, cargo control, machinery control and also administrative systems on a ship.

intelligent terminal A *computer* terminal with its own processor and memory. It will perform simple tasks locally but is linked to a mainframe computer for more complex activities.

interactive A method of computer operation where the user is communicating directly with the *computer*.

intercooler A *heat exchanger* which is fitted between stages in an air compressor in order to cool the compressed air.

intercostal Composed of separate parts, i.e. non-continuous.

interface (1) A circuit whereby one part of a *computer* communicates with another. (2) The boundary between two substances, e.g. oil and water.

interlock An arrangement of controls in which those to be operated later are disconnected until the preliminary settings are correct.

intermediate shafting The lengths of shafting between the *tailshaft* and the *engine* or *gearbox*.

internal combustion engine An engine in which combustion occurs within the cylinders and the products of combustion are used to move the piston on its working stroke.

International Association of Classification Societies An organization which promotes consultation between classification societies on matters of common interest.

International Maritime Organization A permanent body established in 1959 under the aegis of the United Nations to deal with maritime matters on an international basis. It is based in London and the governing body is the Assembly, which meets every two years. Safety and the prevention of pollution are its two chief concerns and various conventions have been produced and adopted as law by the member states of IMO. See *Maritime Safety Committee; Maritime Search and Rescue; MARPOL 73/78; Standards of Training, Certification and Watchkeeping for Seafarers, 1978.*

international standard nautical mile A distance of 1852 metres.

international shore connection A standard sized flange, together with nuts, bolts and washers, which has a coupling fitted which is suitable for the ship's hoses. It is to be used to connect up the shore water main in any port to the ship's *fire main* and equipment. See Figure I.1.

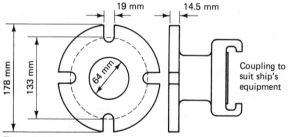

Figure I.1 International shore connection

103

intrinsically safe An item of equipment which cannot release sufficient electrical or thermal energy under any conditions to ignite a particular flammable vapour in its vicinity.

inverter A machine, device or system which converts d.c. power into a.c. power.

involute The curve traced out by a taut string unwinding from a circular section cylinder. It is the basic profile for most modern types of gear teeth.

inward charges The various costs which are incurred by a vessel when entering port.

inward clearing bill A customs document which is issued after declared stores have been checked.

ionization The process, or its result, wherein a neutral atom or molecule acquires a positive or negative charge.

Isherwood system A ship construction system used in oil tankers which consists of continuous longitudinal framing. See *framing systems*.

isolating switch A switch which isolates or separates an electric circuit from the source of power. It is not designed to open when a current is flowing.

isolating valve A *valve* in a piping system which is used to separate a part of the system in an emergency or when maintenance is being done.

isothermal (1) An event occurring at constant temperature, e.g. expansion of a gas. (2) A curve connecting points, or representing quantities measured, at the same temperature.

J

jack A device which is able to lift heavy weights or exert a large thrust in order to move or position equipment. It is usually hydraulically operated.

jack-up drilling unit A mobile offshore *drilling rig* which is designed to operate in water up to about 130 metres deep. The legs of the platform are raised when it is afloat and lowered when it is in use as a drilling rig. See Figure J.1.

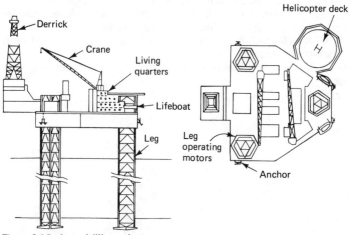

Figure J.1 Jack-up drilling unit

jacket (1) The passageways which surround the cylinders of an engine. They are circulated with fresh water to cool the engine. (2) The support structure of an oil production platform between the base and the deck. (3) Any enclosure which insulates against heat or noise transmission.

jalousie A shutter or cover with louvres. It provides ventilation while restricting the entry of rain and wind.

Jason clause In the event of any accident, damage or disaster before or during a voyage, the owners of cargo or the cargo itself shall contribute with the carrier in *general average* to the cost of any sacrifices that may occur and shall pay *salvage* or any other special charges.

jerk pump A fuel pump which creates a high pressure supply for a short period during the engine cycle.

jerque To search a ship for undeclared goods or to inspect the ship's papers. If no smuggled goods are found a jerque note is issued by the custom authorities.

105

jetsam Goods which have been thrown overboard.

jettison Goods which have been deliberately thrown overboard, or jettisoned. If the action was taken in order to lighten a ship which had run aground it would be a *general average* act.

jetty A structure which projects into the water to a reasonable depth of water. It enables ships or boats to come alongside to load or discharge cargo or passengers. It may also be called a *pier*.

jib The inclined member of a lifting device. See *derrick*.

joint (1) A piece of compressible material which is shaped to fit between two surfaces in order to create a water or gas-tight seal. See *gasket*. (2) A permanent connection between two lengths of cable.

journal The part of a shaft which is supported by a bearing.

junction box A container, with a detachable cover, which provides terminals for the joining of different lengths of cable.

junked well A *well* which has been abandoned due to a blockage which is too expensive to remove.

jury An item which is a makeshift or temporary replacement, e.g. jury rudder.

K

keel A horizontal steel plate of increased thickness which runs along the centreline for the complete length of the bottom plating. It may also mean the complete structure of keel plate, centre girder and centreline strake of *double bottom* plating, where fitted. See Figure K.1.

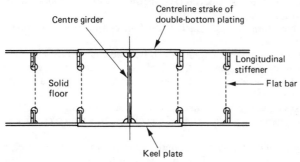

Figure K.1 Keel

keelson Another name for the *centre girder*.

keep A cover for a housing which can be removed, e.g. bearing keep.

kelly A hollow pipe, of square or hexagonal cross-section, which is suspended from the *drill string* and turned by the *rotary table* during drilling.

kelvin scale A scale of temperature, using Celsius degrees, where zero is the theoretical minimum temperature possible for a substance. The *triple point of water* is 273.16 on this scale. A temperature in degrees Celsius is converted by the addition of 273 to give a value on the kelvin scale.

kerosine A distillate of *petroleum* which ranks in volatility between gasoline and gas oil. It is used for lighting, heating and as a fuel in certain types of internal combustion engines.

key A machined metal bar which is used to connect a component to a shaft, e.g. a pulley. It may be rectangular, circular or semicircular in cross section.

keyless propeller A *propeller* which is forced onto the *tailshaft* and grips by friction.

keyway The groove or slot in which a key fits. It must be carefully designed to avoid weakening the shaft or creating an area of stress concentration.

kick A sudden rise in drilling fluid pressure due to a higher pressure in the rock formation. If it is not controlled a *blowout* may occur.

107

kill a well To fill a well with drilling fluid of sufficient density that it will stop the flow of oil.

kinematic viscosity The absolute *viscosity* of a fluid divided by its density at the temperature of viscosity measurement.

kinetic control system A *control system,* the purpose of which is to control the displacement, or the velocity, or the acceleration, or any higher time derivative of the position of the controlled device.

kingpost See *samson post.*

kitchen rudder A two part tube which shrouds the *propeller* and is hinged about a vertical axis. The tube may move or the two halves may move in relation to one another. A steering action, in addition to variation of thrust, is possible.

klaxon An audible alarm device which is used in conjunction with many control systems.

knocking A noise from a diesel engine which may be characteristic at slow speeds or when starting from cold. If it occurs at normal speeds and temperatures it may be due to faulty combustion or a mechanical defect such as too much clearance.

knot A unit of speed which is one nautical mile per hour.

knuckle An abrupt change in direction of a ship's plating which creates an edge.

Kort nozzle A shroud or duct fitted around a *propeller* in order to increase thrust at low speeds. It is often fitted to tugs and trawlers.

L

labyrinth gland A *mechanical seal* which is fitted on turbine rotors to minimise the escape of steam. A series of rings projecting from the rotor and the casing combine to produce a maze of winding passages or labyrinth.

lagging An insulating material applied to surfaces in order to reduce heat transfer, e.g. pipes, or the boiler casing.

lamellar tearing A brittle cracking in steel plate as a result of tensile stresses at right-angles to the plate. It is caused by the contraction of weld metal when cooling.

laminar flow A fluid flow in which the adjacent layers do not mix. In a straight tube the particles of fluid move in straight lines parallel to the axis of the tube. It occurs at low *Reynolds numbers*.

Lanchester balancer A balancing device for certain four-stroke engines which have a uniform crank sequence. It utilizes two shafts, each carrying a balance weight and suitably phased, which are rotated in opposite directions.

land (1) Additional surface on a valve disc or plug, e.g. a *spool valve,* which can be used to vary the flow characteristic of the fluid. (2) To set foot on shore or to bring goods ashore.

language The form in which instructions for a computer are written, e.g. *FORTRAN.*

lapping A polishing process which usually describes the final stage of grinding in a valve and seat when lapping paste is used.

large angle stability The *stability* of a vessel when the angle of inclination is greater than four or five degrees. The criterion for measuring large angle stability is the *righting lever*, GZ.

laser Light Amplification by Stimulated Emission of Radiation. A concentrated beam of monochromatic light which can be used, for example, in cutting and welding processes.

LASH Lighter Aboard SHip. A large vessel which carries up to 80 standard barges each holding up to 800 tonnes of cargo. An on-board crane is used to discharge the vessel which operates into ports with minimum facilities.

latent heat The heat energy required to bring about a change of state of a unit mass of a substance, without change in temperature, e.g. from a solid to a liquid.

launch To move a ship from a position on land into the water until it is afloat.

launching ways The various timber structures upon which a ship is built and from which it will be launched.

Law of Comparison See *Froude's Law of Comparison.*

lay barge See *pipe-laying vessel.*

109

lay days The days permitted for loading and discharge of a vessel according to the *charter party* agreement.

lay up To take out of service for a considerable period.

layout The arrangement of plant and equipment in, for example, a machinery space, or a drawing thereof.

lead (1) A conductor which connects a winding to some terminal point in an electric circuit. (2) A soft, high density, bluish-grey metal with a high resistance to corrosion. It is used as a cable sheath, an electrode in batteries and a constituent of bearing metals. (3) The amount in degrees by which one operation precedes another or a reference point.

lead acid battery A *battery* comprising cells in which the positive electrode is lead peroxide, the negative electrode is lead and the electrolyte is dilute sulphuric acid. A battery of six cells in series produces 12 volts.

leak detector lamp A butane/propane powered gas lamp which is used to check for refrigerant leaks. If a *freon* refrigerant is drawn into the flame it will change colour.

lee The side which is sheltered from the wind.

leg The vertical column of a *jack-up drilling unit*.

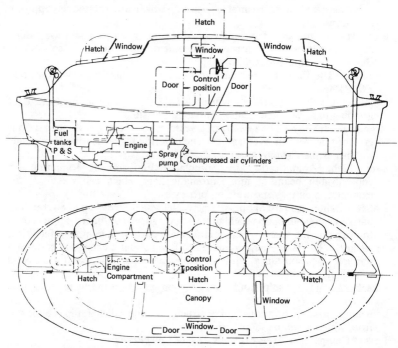

Figure L.1 A lifeboat

110

length between perpendiculars The distance between the *forward* and *after perpendiculars* of a ship, measured along the summer load line. See Figure P.7.

length on waterline The length of a vessel measured along the waterline from forward to aft.

length overall The distance between the extreme points of a ship, forward and aft. See Figure P.7.

less than container load A small quantity of cargo which is insufficient to fill a container.

level compounded See *flat compounded*.

Liberty ship A mass produced type of ship which was built in the United States from 1942 to 1945. It was used as a tramp ship up until the 1970s.

licence A permit to engage in particular activities related to offshore oil, issued by a Government.

lien The right in law to retain goods until any debt relating to the goods has been settled.

lifeboat A rigid vessel secured into *davits* such that it can be launched over the ship's side in an emergency. It is very seaworthy with buoyancy chambers and is often fully enclosed. It is stocked with provisions and equipment to enable survival of the occupants for a reasonable period. See Figure L.1.

liferaft An inflatable rubber vessel of circular shape which is stored, deflated, in a cylindrical container. It is secured in a stand to the deck of a ship. A *hydrostatic release* is used in order that it may float free if the ship sinks. It is stocked with survival equipment similar to that of a lifeboat. See Figure L.2.

lift The distance moved by a valve disc or plug when the valve is fully open.

light crude *Crude oil* which contains a high proportion of lighter fractions and is particularly suitable for gasoline and feedstock production.

light dues A fee paid by a ship in port. It is a contribution to the cost of maintaining lighthouses and other aids to navigation.

light-emitting diode A semiconductor diode which radiates light in the visible region when energized. A red display is usually given.

lighter A barge, usually with a cargo handling derrick, which has no means of propulsion. It is used for cargo transport within port limits.

lighterage A charge made for the use of a barge or lighter.

lightmass The mass of a ship, in tonnes, complete and ready for sea but without crew, passengers, stores, fuel or cargo on board.

lightweight See *lightmass*.

lignum vitae A hardwood which is used as a lining for stern bearings. It can be lubricated by sea water but is subject to some swelling.

limber hole A drain hole usually made in vertical structure at the bottom of a tank to allow the liquid to drain towards the suction well.

line throwing appliance A gun or rocket device which can project a light line between vessels.

linear programming The writing of a computer program or the arranging of

111

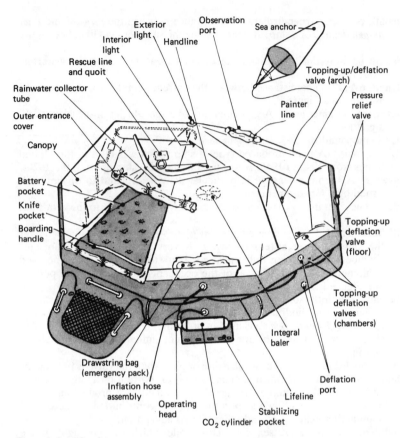

Figure L.2 A liferaft

Labels (clockwise from top): Observation port, Sea anchor, Topping-up/deflation valve (arch), Pressure relief valve, Painter line, Topping-up deflation valve (floor), Topping-up deflation valves (chambers), Deflation port, Lifeline, Stabilizing pocket, CO₂ cylinder, Operating head, Inflation hose assembly, Drawstring bag (emergency pack), Boarding handle, Knife pocket, Battery pocket, Canopy, Outer entrance cover, Rainwater collector tube, Rescue line and quoit, Interior light, Exterior light, Handline, Integral baler

a sequence of operations in order to achieve an optimum result.

linearity A measure of the maximum deviation from a linear input-output relationship, usually expressed as a percentage of the full scale of measurement.

liner (1) A replaceable metal cylinder which forms the cylinder in an internal combustion engine. (2) A cargo or passenger carrying vessel which makes regular, scheduled voyages.

liner pipe A steel pipe which is suspended within the *casing* of a well.

lines fairing The process of checking for correspondence between the various drawings of the lines plan to ensure that all curved surfaces are 'fair', i.e. run smoothly and evenly.

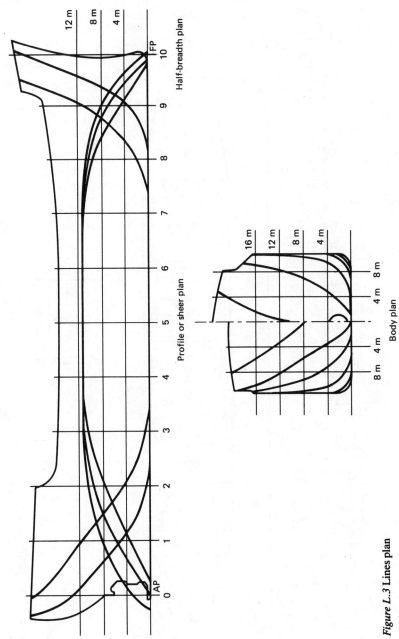

Figure L.3 Lines plan

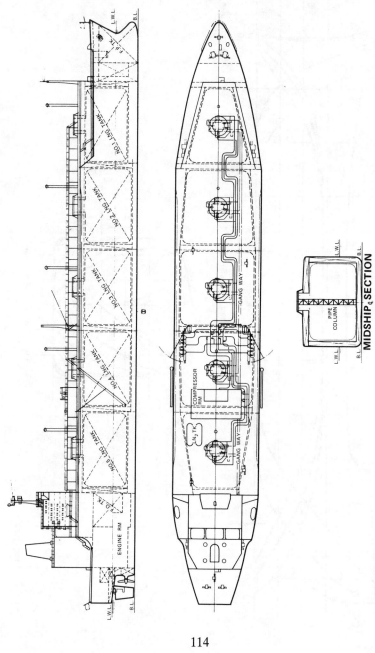

Figure L.4 Liquefied gas tanker

114

lines plan A scale drawing of the moulded dimensions of a ship in plan, profile and section. The individual views are known as the *half-breadth plan*, the *profile* or sheer plan and the *body plan*. See Figure L.3.

lip seal A shaft seal used to prevent the entry of sea water or the loss of oil from a *stern tube bearing*. A shaped circular rubber ring is held against the rotating shaft by springs and the existing pressure differential.

liquefaction To change into a liquid state. It usually refers to the conversion of a gas into a liquid which requires the gas to be cooled below its critical temperature and sometimes also compressed.

liquefied gas tanker A vessel which has been specially designed and constructed for the carriage of various natural gases in liquefied form. Different tank systems exist together with arrangements for pressurizing and refrigerating the gas. See Figure L.4.

liquefied natural gas A naturally occurring gas which is 75–95% methane and has a boiling point of $-162°C$ at atmospheric pressure. It has a critical temperature of $-82°C$ and cannot, therefore be liquefied at normal temperatures. It is usually carried in liquefied form at atmospheric pressure and a temperature of $-164°C$.

liquefied petroleum gas This may be *propane,* propylene, *butane* or a mixture of each. All have critical temperatures above ambient and can be liquefied at low temperatures at atmospheric pressure, normal temperatures under pressure or some condition in-between.

liquid crystal display Liquid crystal cells whose light transmitting properties vary with the applied electric field. Numbers or letters can be displayed by the use of an array of straight lines.

liquid receiver A vessel, fitted in a refrigeration system, to store a reserve of refrigerant or to hold the complete charge during maintenance.

list See *heel* (1).

live A circuit or device which is connected to a source of voltage or electrically charged.

Lloyd's Corporation An insurance underwriting organization which is based in Lime Street, London, UK.

Lloyd's List A daily newspaper which is published by *Lloyd's Corporation.* It lists movements of ships, casualties and other shipping matters.

Lloyd's Machinery Certificate A notation awarded to ships where the machinery has been built according to the Classification Society's rules and satisfactorily proved on sea trials.

Lloyd's Register of Shipping A *Classification Society* which produces its own 'Rules and Regulations for the Classification of Ships'. The ship and all its machinery when built and tested according to the rules will be awarded various appropriate *class notations*. Regular inspections or *surveys* are required in order to remain classified.

Lloyd's Signal Station An establishment which provides information to *Lloyd's Corporation* regarding the movement of ships within its area or region.

Lloyd's Surveyor A qualified inspector who ensures compliance with classification rules by attendance during construction, repairs and maintenance, throughout the life of a classed ship.

load In a *kinetic control system* this may be the controlled device, or the properties, e.g. inertia, friction, of the controlled device that affect the operation of the system. For a process control or regulating system, this is the rate at which material or energy is fed into, or removed from, a plant.

Load Line Rules The *Merchant Shipping (Load Line) Rules, 1968.* Regulations based on the IMO Load Line Convention, 1966, which set out the requirements for a minimum freeeboard which must be indicated on a ship's side by a special load line mark.

load lines A series of lines or marks which are situated forward of the load line mark on either side of a ship. They denote the minimum freeboards within certain geographical zones or in fresh water. See Figure L.5.

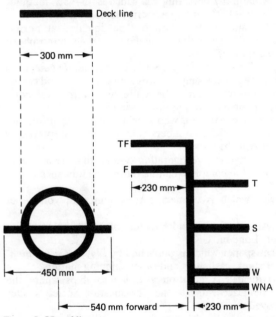

Figure L.5 Load lines

load-on-top The loading of a new cargo of crude oil on top of oil which has been recovered from tank cleaning operations.

lock A region of a canal or river which can be enclosed by gates. It is then filled or emptied in order to raise or lower ships whilst afloat.

locked rotor torque The minimum torque produced by an a.c. motor when the rotor is locked and the rated voltage is applied, e.g. at the instant of starting.

116

locking nut An additional nut or a special type of nut which is used to prevent loosening of a connection due to vibration or any other cause.

locking pintle A *pintle* which has a shoulder of increased thickness at its lower end which prevents excessive lifting of the rudder.

lofting See *mould loft.*

log A device which measures the distance travelled and also the speed of a ship moving through the water. See *pitometer.*

log book A book which the Master of a ship is required to complete during a voyage with certain specified information.

logic (1) Science of step-by-step formal reasoning. (2) A method of operation using two-state, i.e. on or off, devices which are used in logic devices to perform arithmetic operations in a computer or a step-by-step sequential control operation.

logic devices Elements which combine two-state operations in various ways in order to provide particular functions such as AND, OR, NOT. They are also known as logic *gates.*

logistics The science and practice of supply and provisioning related originally to troops but now applied more universally.

loll See *angle of loll.*

long ton A unit of mass of 2240 pounds.

longitudinal framing A *framing system* which is often adopted for tankers and must be used for vessels greater than 198 metres in length. Longitudinal stiffeners are used along the ship's sides and throughout the tank length. Side and bottom transverses are used to support the longitudinals against compressive loading. See Figure F.4.

loop scavenging A *scavenging* process in a two-stroke engine where the air enters through inlet ports and exhaust gas leaves through exhaust ports which are one above the other. The scavenging air travels in a complete loop.

loran LOng RAnge Navigation. A radio navigation system which uses transmissions from shore stations in order to determine a position on a special chart.

lost motion clutch A servomotor arrangement which moves the camshaft through the angular period between top dead centre in one direction to the top dead centre position in the other direction, e.g. when directly reversing a slow-speed diesel engine. See Figure L.6.

lower calorific value See *calorific value.*

lubricating oil A high boiling point product of the crude oil refining process. The various properties required are obtained by blending and the introduction of additives such as oxidation inhibitors, dispersants and detergents.

lubrication The process of minimizing friction and wear between moving metal parts by the formation of a film of oil between them.

lubricator quill A small bore pipe which passes through the jacket water cooling space or a sleeve therein and fastens into the cylinder liner. It is used to inject cylinder lubricating oil between the piston and cylinder.

117

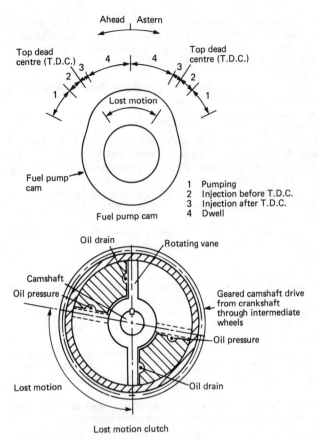

Figure L.6 Lost motion clutch: (a) fuel pump cam, (b) lost motion clutch

luff (1) The side exposed to the wind. The opposite of lee. (2) To raise or lower a derrick or the jib of a crane.

lug A projection from a casting or a structure.

Lutine bell The bell from HMS Lutine which hangs in the headquarters of *Lloyd's Corporation.* It is rung whenever important announcements are made to members.

118

M

machine language A set of binary instructions that can be interpreted or executed directly by a *computer*.

machinery rating See *rating*.

machinery space The compartments which house the main and auxiliary machinery.

magnetic crack detection A *non-destructive testing* technique using a mixture of iron filings in thin white paint which is spread over the surface to be examined. A large magnet is then attached over the area of interest and any discontinuities show up as concentrations of iron filings.

magnetic disk A thin circular plate with magnetic surfaces upon which data can be written by a *computer*. It is also known as a floppy disk.

magnetometer An instrument which measures the strength and direction of a magnetic force. It is used in geological surveys.

Magnus effect The development of a lifting force normal to the axis of rotation and the stream flow by the high speed rotation of a cylinder in a fluid stream. It has been applied in ship propulsion and also for rudders.

maiden voyage The first voyage of a new ship after acceptance by the owner.

maierform A bow design with considerable rake.

main circuit breaker A *circuit breaker* fitted to the main switchboard.

main deck The uppermost continuous deck which is used in structural calculations to ensure adequate longitudinal strength and resistance to bending.

main hatch The *hatch* which is used for the heaviest cargo. The *Ship's Official Number* and *Registered Tonnage* are cut into its coaming.

main inlet The entry point for sea water which supplies the main *circulating pumps*. A *sea inlet box* or *sea tube* will be fitted into the ship's side plating at this point.

main steam stop valve A *valve* which is fitted in the steam supply line from a boiler. It is usually of the non-return type.

main vertical fire zones Those sections into which the hull, superstructure and deckhouses are divided by *A-Class divisions,* the mean length of which must not exceed 40 metres.

maintenance Any action which is carried out to return or restore an item to an acceptable standard.

maintenance costs The costs incurred in keeping equipment in an operating condition.

make-up feed The *feed water* used to replace losses in a feed system.

manganese A metal element which is used in the steel making process and is also added to improve the mechanical properties of *steel*.

manganese bronze A high tensile strength alloy of iron, tin, aluminium and manganese. It is used for valve bodies, propellers and other applications where corrosion resistance is required.

manhole An oval opening to enable access into a boiler drum, a double bottom tank or other enclosed space.

manifest A document which gives a complete list and description of the cargo carried on a ship.

manifold A chamber which acts as a connection point for various valves and pipelines.

manifold centre The grouping point for production pipelines from various wells into one or more pipelines for transfer ashore.

manning The number of personnel who form the crew of a ship.

manning scale The minimum number of qualified persons who must form part of a ship's crew according to law.

manoeuvring valves The valves which admit steam to the ahead or astern turbines. There are three, the ahead, the astern and the guarding or *guardian valve*.

manometer A pressure measuring instrument which balances the applied pressure with a column of liquid. The height of the column of liquid is then a measure of the applied pressure.

margin line An imaginary line drawn 75 mm below the bulkhead deck at the ship's side. If in any damaged condition the vessel were to float at a waterline tangential to the margin line this is considered as the limit of flooding without sinking.

margin plate A sloping plate which extends the length of the bilge and acts as a cover for the double bottom tank and a collecting bay for the bilge.

marine insurance A contract which indemnifies the owner of a ship or its cargo against losses related to a marine adventure.

marine riser A large diameter tube which extends from the *blowout preventer* on the sea bed, to the *drilling platform*. It is used as a return path for the drilling fluid.

marinization The modification of machinery or equipment in order to make it suitable for use and operation at sea.

maritime court A court which deals with any legal action or an official enquiry relating to ships or shipping.

maritime lien A *lien* placed on a ship, its equipment or cargo.

Maritime Safety Committee The most senior of the technical committees of the *International Maritime Organization*.

Maritime Search and Rescue An *International Maritime Organization* Convention which is intended to improve the existing arrangements for carrying out search and rescue operations following accidents at sea.

markings The required marks on a British ship which include the name on each side of the bow and the stern, port of registry on the stern, draught marks forward and aft on each side, the official number and the loadline markings.

MARPOL 73/78 The 1973 Marine Pollution Convention and the 1978 Protocol which in effect absorbed the parent convention. They lay down numerous requirements to minimize or limit the affects of pollution from vessels.

mast A tubular steel erection which carries various items of navigational equipment and fittings, e.g. lights, radar, etc.

mast table A small platform at the base of a mast which supports the hinged heel bearings of derricks.

master The officer in command of a merchant ship.

master controller A controller which is used in a *cascade control system*. It provides an output which acts as a variable desired value for a *slave controller*.

mat supported jack-up A type of *drilling rig* which rests on the sea bed on a large mat or frame.

mathematical model The use of mathematical equations to express the relationships between variable quantities in a system. These equations are then appropriately combined to form a single equation which is the mathematical model.

maximum continuous service rating The practical limit of output for a diesel engine which is to be run continuously. See *rating*.

mayday A radio distress call.

mean effective pressure The mean or average gas pressure considered to act during the power event of an internal combustion engine. It is used to determine the work done in the engine cylinder. The value is an indication of the loading on the moving parts by combustion.

mean time between failures The ratio of the accumulated operating time for a sample of parts to the total number of failures in the sample, for specified conditions of operation.

means of escape A route which provides a means of escape during a fire or other emergencies from, for example, a control room or a shaft tunnel.

measured mile A distance of one nautical mile marked on a region of shore by two posts at each end. A ship will use these markers for speed measurements when on *sea trials*.

measuring element The element which receives the signal from the detecting element and provides a signal representative of the controlled condition.

mechanical efficiency The ratio of *shaft power* to *indicated power* for an internal combustion engine.

mechanical planer A machine in which steel plate may be planed or cut to size using roller shears. Milling heads are used for accurate edge preparation.

mechanical seal A shaft sealing arrangement fitted to pumps in place of a *stuffing box* and gland packing assembly.

medium speed An internal combustion engine operating in the range 250 rev/min to about 750 rev/min.

medium voltage Voltage in the range 250 up to 650 volts.

Megger tester See *insulation resistance*.

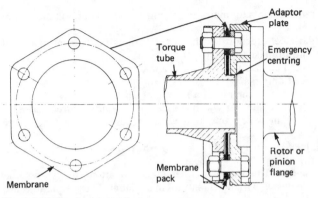

Figure M.1 Membrane coupling

membrane coupling   A form of *flexible coupling* which uses packs of thin membrane plates as the flexible element. See Figure M.1.

membrane tank   A containment tank for liquefied gases which consists of thin metal plate. It is surrounded and supported by insulation which is load bearing. Single and double membrane designs exist with the insulation between the membranes in the double type.

membrane wall   A *waterwall tube* arrangement in a boiler furnace where the tubes have a metal strip welded between them to form a completely gas-tight enclosure. It is also known as *monowall*.

memory   The data storage part of a *computer*.

Mercantile Marine   Another term for *Merchant Navy*.

Mercantile Marine Office   The official office where seamen may sign on and off ships. Various personnel matters related to merchant seamen are handled here.

Merchant Navy   The ships and crews of the merchant fleet.

Merchant Shipping Acts   The Acts of Parliament relating to the Merchant Service. The first Act was in 1894 and many supplementary or amending Acts have been made.

Merchant Shipping (Load Line) Rules 1968   See *Load Line Rules*.

mesh   (1) The movement of gear wheels between engaging and disengaging. (2) The space between metal strands of a gauze.

metacentre   The point where a vertical line through the centre of buoyancy of an inclined ship intersects the vertical line through the centre of gravity when it is floating in equilibrium. This is the *transverse metacentre* and a longitudinal metacentre can be determined in a similar manner. See Figure M.2.

metacentric height   The distance between the *centre of gravity*, G, and the *metacentre*, M, of a ship. It is often referred to as the GM and for stability must be a positive value, i.e. G lies below M. See Figure M.2.

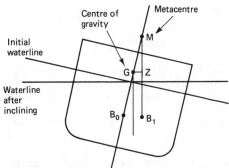

Figure M.2 Metacentric height. B_0 = centre of buoyancy before inclining; B_1 = centre of buoyancy after inclining

metal inert gas welding An *arc welding* system which uses an inert gas to shield the bare wire *electrode*. The electrode and the gas are fed through the torch.

meteorology The study of atmospheric phenomena as a means of forecasting the weather.

methane A colourless gas, obtained from petroleum, and the principal constituent of natural gas. It has a critical temperature of −82°C and a boiling point of −162°C at atmospheric pressure. It is usually liquefied and carried at atmospheric pressure and a temperature of −164°C. It is used as an industrial and domestic fuel and a feedstock for making ethylene and *acetylene.*

method study A branch of *work study* which examines the ways of doing work in order to develop simpler and more cost effective methods of production.

metre The *SI unit* of length. It is defined as 1 650 763.73 vacuum wavelengths of radiation of krypton-86.

metric ton A ton of 2204.6 pounds.

mica A mineral composed of aluminium silicate and other silicates which can be split into thin transparent plates. It is used as an insulating material.

microbial degradation A phenomenon which affects the lubricating oil and coolants of diesel engines. Bacteria, mould or yeast organisms grow under certain conditions and either produce corrosive growth products, create corrosive dispersions of oil and water, change viscosity, reduce the effectiveness of additives or promote *electrochemical corrosion.*

microcomputer A digital *computer* comprised of a *microprocessor* and an electronic *memory.*

micrometer A precision measuring instrument for measuring small distances.

micrometre (formerly *micron*) One millionth of a *metre.*

123

microprocessor A *central processing unit,* for a *computer,* located on a single silicon chip.

midship section The transverse section at amidships.

midships See *amidships.*

mild steel An alloy of iron and 0.15–0.23% carbon with various other metals to improve the properties. Classification societies designate five grades, A–E, which must be manufactured, inspected and tested in order to be approved.

mill scale An oxide of iron which forms on the metal surface during manufacture. It is usually removed by suitable surface treatment prior to priming and painting.

mimic diagram A line diagram of a pipe system or an item of equipment which includes miniature alarm lights or operating buttons for the relevant point or item in the system.

mineral insulated copper covered cable An electric cable which uses magnesium oxide powder as insulation and has a copper sheath. It is considered reasonably fireproof.

mineral oil A general term for petroleum and other hydrocarbon oils which have been obtained from the Earth. It may refer particularly to *lubricating oils.*

miniature circuit breaker A very small air *circuit breaker* which is fitted into a moulded plastic case and has a current rating of 5–100 amps. It is often used in a final distribution board instead of a fuse as it is fitted with a thermal overload trip.

Ministry of Transport See *Department of Transport.*

misalignment (1) The *deviation* present in a position control system. (2) The distance between the axes of two shafts which are to be coupled together.

mobile drilling unit A *platform* or *drill ship* which may move from one drilling location to another.

model A scale reproduction of the hull of a vessel which is tested in a *towing tank* in order to determine resistance and hence powering required by the full size ship.

modem A device which connects a *computer* with a telephone line to enable transmission of data to another similarly connected computer.

modulation The use of a *carrier wave* to transmit a signal. Some characteristic of the carrier wave is changed by the signal. See *frequency modulation.*

module (1) The pitch diameter of a gear wheel divided by the number of teeth. (2) An assembly of equipment and piping which is installed as a complete unit into the machinery space of a ship. (3) A prefabricated component of an offshore oil rig, e.g. accommodation module.

modulus of elasticity The ratio of *stress* to *strain* for a material, within the range of elasticity. It has the units of stress.

molybdenum A heavy metal element which is added to high speed steels. As molybdenum disulphide it is added to industrial lubricants which must withstand high temperatures and pressures.

moment causing unit trim The moment acting on a ship which will change its trim by one unit. If Imperial units are used this is the moment to change trim by one inch, MCT 1 in. If SI units are used this is the moment to change trim by one metre (although centimetre is sometimes used), MCT 1 m.

moment of inertia The product of the mass and the square of its perpendicular distance from the axis considered, which is summed for all the elements in the body considered.

momentum theory A theory of action of the marine propeller where it is considered as a means of accelerating water.

monel metal An alloy of 68% nickel, 29% copper and small quantities of iron, manganese, silicon and carbon. It has a high resistance to corrosion and is used for condenser tubes, pump fittings and propellers.

monkey island The navigating position above the wheelhouse. It is sometimes known as the flying bridge.

monowall See *membrane wall*.

moonpool An open area on a *drilling rig* through which the drilling operation takes place.

moor To secure a ship by attaching to a buoy, a position ashore or by anchoring.

mooring winch A *winch* with a barrel or drum which is used for hauling in or letting go the mooring wires. A warp end is also fitted to assist in moving the ship.

motion compensator A mechanism which is used by mechanical equipment on a floating drilling rig, e.g. a crane, to counteract any movement due to the rig's motion in the sea.

motor element The element which moves the correcting element as a result of a signal from an automatic controller.

motor enclosures A casing provided according to the location of the motor, e.g. *flameproof, drip-proof, hose proof.*

motor starter An electric controller for starting a motor from rest, accelerating it up to normal running speed and also stopping it.

motor vessel A ship propelled by a diesel engine.

mould loft A large covered area in a shipyard, which has a wooden floor. The ships lines are drawn out on the floor to full, or some smaller scale, size.

moulded breadth See *breadth moulded*.

moulded case circuit breaker A small, compact, air *circuit breaker* in a moulded plastic case. It has a current rating of 30–1500 amps and various overload trips built-in.

moulded depth See *depth moulded*.

moulded draught The distance from the summer load line to the base line, measured at the midship section.

mouse A hand-held device which is connected to a *computer* and can, with certain programs, be used to manipulate screen displays without reference to the keyboard.

moving coil meter An instrument which is used to provide a measurement of voltage or current. It consists of a coil wound on a soft iron cylinder which is free to move within a radial magnetic field. A flow of current will create a force which moves a needle over a scale. See Figure M.3.

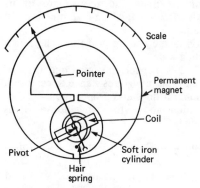

Figure M.3 Moving coil meter

moving iron meter An instrument which is used to provide a measurement of voltage or current. It consists of a fixed coil which carries the current to be measured and causes movement of a pivoted piece of soft iron.

mud See *drilling fluid*.

mud box A coarse strainer fitted into the machinery space bilge piping. A straight *tailpipe* is led from the mud box to the bilge. It may also be called a *strum box*.

muff coupling (1) A solid *coupling* using two sleeves which are taper fits one into the other. The inner sleeve fits over the two shafts and a friction grip is created when the outer sleeve is forced onto the inner sleeve. (2) A *flexible coupling* between a steam turbine and gearbox. A boss with projecting gear teeth is attached to each shaft and a muff with inwardly projecting teeth is fitted over the two bosses. There is a small clearance between the teeth to provide flexibility.

multi-hull ships *Catamaran* or occasionally trimaran vessels with two or three hulls respectively. They are placed parallel to one another and some distance apart, connected by a bridge structure.

multimeter An electrical measuring instrument which measures current, voltage and resistance over a variety of ranges. It will measure d.c. or a.c. values.

multiplexer A selector which connects various inputs, one at a time, to a common output.

multivibrator An *oscillator* which produces a repetitive pulse or rectangular wave form.

126

N

naphtha A product of the distillation process which is an oil heavier than gasoline and lighter than kerosine. It is used as a feedstock for ethylene and also aromatics.

natural depletion The gradual exhausting of an oil or gas reservoir where the naturally occurring gas or water forces have been used as the driving mechanism.

natural frequency The *frequency* at which free oscillation occurs.

natural gas A mixture of *methane, ethane, propane, butane* and pentane which is found and released as a result of oil drilling operations. It is about 95% methane and has a boiling point of −162°C.

nautical Relating to ships, their crew or navigation.

nautical mile A distance of 1852 metres.

navigation To direct the course of a ship by the use of astronomical and mathematical means of determining position.

navigation lights The various lights which a ship at sea, under various circumstances, must show to comply with the International Regulations for Preventing Collisions at Sea (1972). Their position, colour, angle and range of visibility are detailed in the regulations.

navigator A qualified person who directs the course of a ship and determines its position.

neoprene A trade name for *polychloroprene* (PCP), which is a sheath material for electric cables. It has largely been superseded by *hypalon*.

nest of tubes An assembly of parallel tubes, with a tube plate at one or both ends, which is used in *heat exchangers*.

nested plates The arrangement of oddly shaped plates, which are to be cut from a large plate, in such a way as to minimize wasted material.

net positive suction head The difference between the absolute pump inlet pressure and the vapour pressure of the liquid. It is expressed in metres of liquid. See Figure H.2.

net tonnage A measure of the useful capacity of a ship. It has previously been found by deducting the volume of spaces necessary for the propulsion and operation of the ship, e.g. accommodation, equipment and machinery spaces, from the gross tonnage. The International Convention on Tonnage Measurement of Ships, 1969, requires the use of a formula related to the volume of cargo carrying spaces.

network analysis An *operational research* technique which uses a network or arrow diagram to represent a planned programme. See *critical path method* and *programme evaluation and review technique.*

neutral The point at which a polyphase star connected system of windings are joined together.

neutral axis The geometrical centroid of a section. For a beam (or a ship) subjected to longitudinal bending, this is the line of zero stress which lies across a transverse section of the longitudinal plane.

neutral earthing resistor A resistor fitted in the earth path of a high voltage (more than 3.3kV) system. Its value is chosen to limit the maximum earth fault current to not more than the generator full load current.

neutral for a.c. systems The return conductor, in a three-phase star connected system, which is considered to be almost at earth potential, where no earth exists.

newton The *SI unit* of force which is one kilogram metre per second per second.

nickel A silver-white metallic element which is used as a surface coating and as an alloy with *steel* to improve mechanical properties. When alloyed with copper it produces *cupro-nickel* which has a good resistance to corrosion.

nitriding A surface hardening process for special steels. The steel is heated to about 500°C in an atmosphere of ammonia gas for many hours.

nitrile rubber A synthetic rubber which is a copolymer of butadiene and acrylonitrile. It has a good resistance to petroleum based substances and is often used in seals and jointing materials with these materials.

nitrogen A colourless, odourless, gaseous element which occupies about 80% of the atmosphere. It is an inert gas.

No cure–No pay The principle of payment for salvage under the terms of the Lloyd's Standard Form of Salvage Agreement.

node A point which is at rest in a vibrating body.

noise (1) A sound which is irritating and unwanted. The ability to hear a noise is related to certain frequency ranges. Sound level is measured in *decibels* above an arbitrary zero. Continued exposure to high sound levels can cause permanent hearing damage. (2) Any interference in a communication or transmission system.

non-associated natural gas An accumulation of gas which exists without any oil being present.

non-combustible material A material which neither burns nor gives off flammable vapours in a sufficient quantity to self-ignite when heated to 750°C in an approved test.

non-destructive testing Various tests which do not damage the material under test and can, therefore, be used on the finished item, if required. Testing can be done for properties such as hardness, quality and soundness. See *dye penetrant testing, radiography, ultrasonic testing, ultra-violet light crack detection.*

non-return valve A *valve* which is designed to prevent reverse flow. Where the valve disc is not attached to the spindle it is called a screw down non-return (SDNR) valve. It may not have a spindle in which case it cannot be closed. See Figure N.1.

non-volatile memory A permanent *memory* whose contents remain even when power is disconnected from the computer.

normal circulation The circulation of drilling fluid through the drill pipe

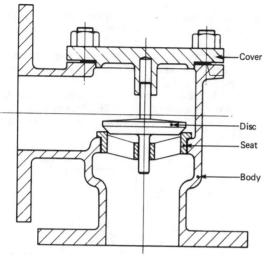

Figure N.1 Non-return valve

and bit, up through the annulus between the drill pipe and the bore wall to the surface and into the mud pit.

normalizing A heat treatment for *steel* where it is heated to a temperature in the region of 850–950°C depending on its carbon content, and then allowed to cool in air. A hard strong steel with a refined grain structure is produced.

Not Under Command A description of a ship which is unable to manoeuvre due to some cause, e.g. steering gear failure. It must display appropriate signals to warn other vessels.

notation See *class notation*.

notch A small opening cut in a structural steel plate, usually to permit the passage of continuous material, e.g. a longitudinal stiffener.

notch ductility The property of a material whereby it will withstand stress due to a notch or area of stress concentration without cracking or failure.

notch tough steel *Steel* which has good notch ductility.

Notice of Abandonment A formal statement given by the assured, following a loss, that he intends to abandon his interest and claim for a *constructive total loss* from the *underwriter*.

Notice of Readiness A formal statement by a shipowner to a charterer that the ship is ready to load.

Notice to Mariners An official notice giving details of changes relating to aids to navigation, charts or other matters of importance to navigators.

Noting Protest A sworn statement on a matter relating to a protest which is usually made before the full facts are known as a means of registering the matter.

nozzle A narrow passage which is shaped so as to convert pressure energy into kinetic energy (a high velocity jet), e.g. steam which is to be used in a steam turbine.

nozzle box The enclosed region around a number of first stage or entry nozzles of a steam turbine. Several nozzle boxes will be fitted around the periphery, each with its own nozzle control valve.

nozzle-flapper An arrangement used in many pneumatic devices to transduce a displacement into a pneumatic signal. The flapper is pivoted and moves to close or open a restriction which will vary the air flow through the nozzle. See Figure N.2.

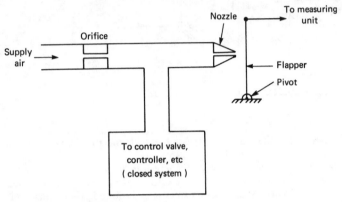

Figure N.2 Nozzle-flapper mechanism

null measurement A means of measurement where a zero deflection of the instrument is established by creating an effect which nullifies or balances that of the quantity being measured.

numerical control The automatic control of a machine by the insertion of numerical data. The numerical data is a sequence of numbers which fully describe a part to be produced.

nylon A synthetic polymer which is chemically inert and resistant to erosion and impingement attack. It is used for orifice plates, valve seats and as a coating for salt water pipes.

O

O-ring A ring of circular cross section made from *neoprene* or a similar material. It is used as a seal to prevent leakage of air, water or oil.

observation tank A feed water tank in an open feed system which enables inspection of drains from oil tank heating coils and other services which may become contaminated with oil.

octane An alkane hydrocarbon which is a natural constituent of *petroleum*. It is a colourless liquid.

Oertz rudder A special design of *flap rudder*.

Off-hire survey See *On-hire survey*.

officer A senior member of a ship's crew, usually the holder of a *Certificate of Competency* as either engineer or deck officer.

official log See *log book*.

offset A continuing *deviation* usually occurring when proportional action is used alone.

offsets Measurements which are taken from a faired *lines plan* to give the coordinates through which the curved lines representing the hull must be drawn.

ohm The *SI unit* of measurement for *resistance* and *impedance*.

ohmmeter An instrument which measures electrical resistance in either ohms or megohms.

oil See *crude oil, fuel oil, lubricating oil.*

oil consumption The quantity of oil, either lubricating or fuel, which is used by an engine in a specific period.

oil cooler A *heat exchanger* which is used to reduce the temperature of lubricating oil.

oil field A geological structure in which an oil accumulation of commercial proportions exists.

oil-in-water monitor A measuring instrument which determines the concentration of oil in discharging water. Where the permitted level is exceeded an alarm is given and the water flow is diverted to a slop tank.

oil mist detector See *crankcase monitoring.*

oil ring A loose ring fitted on a journal and dipping into a reservoir of oil. As the journal rotates the ring moves over it and carries oil up from the reservoir.

oil sand An underground region of sandstone which acts as an oil reservoir.

oil shale A fine grained underground rock which, after distillation, will yield gaseous and oil products.

oil string The final string or section of *casing* in a well which is to be completed in order to become oil producing. It may also be known as an inner conductor or production casing.

oil tanker A vessel which has been designed and constructed for the carriage of oil in tanks, see Figure O.1. The size and location of these tanks is dictated by **MARPOL 73/78**. See *products tanker, very large crude carrier, ultra large crude carrier.*

oil trap   A geological structure or permeability barrier which prevents the movement of oil and thus assists in the formation of an underground oil accumulation.

oil treatment   The preparation of *fuel oil* or *lubricating oil* for use in a diesel engine. It will involve storage and heating to allow separation of any water present, coarse and fine filtering to remove solid particles and also centrifuging to further clean the oil.

oil/water interface detector   A sensor which must be fitted in oil tanker slop tanks to comply with **MARPOL 73/78**. During pumping of the slop tank, it determines the *interface* level at a low position and stops further discharge.

oiltight bulkhead   See *bulkhead.*

oily water separator   An item of equipment which separates oil from oil/water mixtures in order to create a 'clean' water discharge which meets the requirements of international legislation with respect to oil pollution, i.e. **MARPOL 73/78**.

On-hire survey   The examination of a vessel prior to charter to establish the condition of a vessel, contents of tanks, etc. A similar *Off-hire survey* will take place at the end of the charter to ensure a return of the vessel in good condition.

on-off action   A particular case of two-step action, where one of the output signal values is zero.

onstream   The stage in oil production when oil or gas begins to flow.

open charter   A *charter party* which does not specify any particular cargo or ports for the vessel.

open-circuit fault   An electric circuit fault resulting from a break in a conductor such that a complete circuit no longer exists.

open-loop control system   A control system without monitoring feedback.

open sea   Any ocean or sea outside of territorial limits, i.e international waters.

open water efficiency   The ratio of *thrust power* to the power absorbed by the propeller if it were operating without a hull attached, i.e. in open water.

operating company   An oil company or consortium which undertakes the exploitation of oil or gas discoveries.

operating costs   All expenses involved in the continuous operation or availability for operation of machinery, equipment, production lines, a transport system, etc.

operating manual   An instruction book provided by a manufacturer which details the correct operating procedure for a piece of equipment.

operational amplifier   A very high gain directly coupled *amplifier* which is able to generate a variety of different functions, some of which include mathematical operations.

operational research   The use of scientific techniques to obtain quantitative

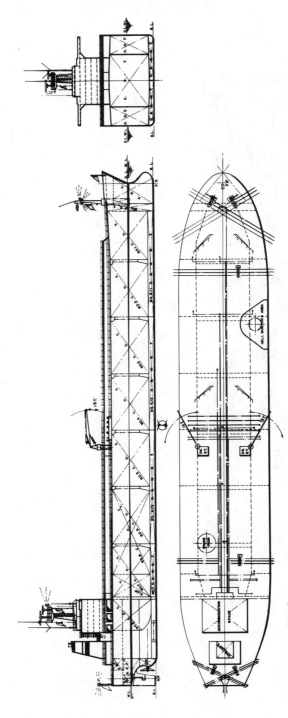

Figure O.1 An oil tanker

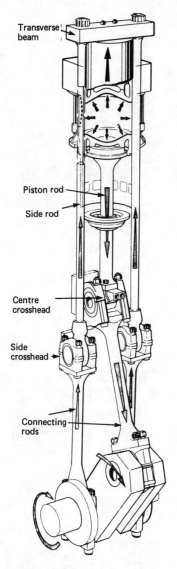

Figure O.2 An opposed piston engine

134

values in order to assist executive decision making. It can also embody an examination of working methods and processes in order to improve *productivity*.

opposed piston engine An engine in which the exploding fuel acts on two pistons simultaneously in the same cylinder. The two pistons are driven in opposite directions and appropriate linkages connect them to a crankshaft which is driven round. See Figure O.2.

optimal The best with respect to a particular set of conditions or parameters.

ore/bulk/oil carrier A combination *bulk carrier* which has been designed and constructed to carry either oil, ore or any dry loose material, e.g. coal, in bulk. It is a single deck vessel with the cargo carrying section divided into holds which also act as tanks and must be appropriately constructed. See Figure O.3.

ore carrier A *bulk carrier* which has been designed and constructed for the carriage of metal ores in bulk. A particular characteristic is the deep double bottom to raise the centre of gravity of the very dense cargo.

ore/oil carrier A combination *bulk carrier* which has been designed and constructed to carry either liquid or solid cargoes in bulk. When ore is carried only the centre tank section is used. All bulkheads and hatches must be oiltight.

organization and methods The study of clerical organization and procedures to promote efficiency in offices.

Organization of Petroleum Exporting Countries A consortium of oil producing countries which develops common policies on oil pricing, production, etc.

orifice plate A thin circular plate with an axial hole. It is fitted into a pipeline in order to enable a differential pressure measurement across it. This differential pressure is then a measure of the liquid flow rate through the pipe.

Orsat apparatus An apparatus which is used to analyse exhaust gases, usually from a boiler. A gas sample is passed through solutions which absorb and enable measurement of carbon dioxide, carbon monoxide and oxygen in turn.

oscillation A sustained periodic variation about some specified reference value.

oscillator An electronic circuit which is designed to produce an output e.m.f. with a particular frequency and waveform.

oscilloscope A measuring or display instrument which uses a cathode ray tube to show the instantaneous values of varying electrical quantities with respect to time or some other quantity.

osmosis The diffusion of a solvent through a semi-permeable membrane, which separates two solutions of different concentrations, such that the concentrations are equalized. See *reverse osmosis*.

Otto cycle engine The operating cycle for a four-stroke engine in which the burning of fuel and the exhaust event are both considered to take place at

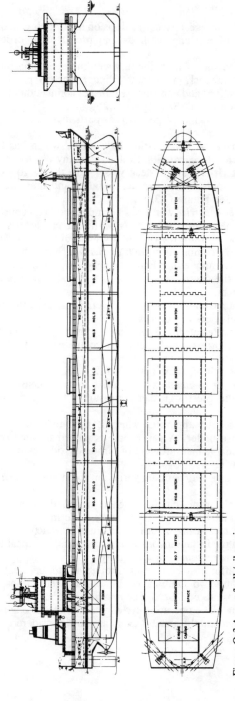

Figure O.3 An ore/bulk/oil carrier

constant volume. The cycle requires two revolutions of the crankshaft.

out of phase diagram See *draw card*.

outboard In a direction away from the centreline of the ship.

outer conductor A short string or length of *casing* which serves to anchor equipment on the sea bed and also acts as a return path for the drilling fluid in the early stages of drilling.

output The results of computer processing of data.

overall (1) A measurement between two extreme points, e.g. length overall for a ship. (2) A one piece garment which covers most of the body, i.e. boiler suit.

overboard To pass from a ship into the water.

overcompounded A *compound wound* d.c. *generator* with a series winding which causes the terminal voltage to increase with load.

overdamping An amount of *damping* in a control system which is greater than that required for critical damping.

overflow pipe A pipe, which is usually fitted to an oil tank, to direct any overflowing oil into another tank to avoid spillage or pollution. The overflow tank receiving this oil must be fitted with a level alarm.

overhaul (1) To strip down or dismantle for examination, adjustment or repair and subsequently reassemble ready for use. (2) To move up level with and overtake.

overlap (1) The crank angle or period when the inlet and exhaust valves of a four-stroke engine are both open. (2) Infused metal lying over the parent metal at a welded joint or seam.

overload (1) A load which is in excess of the rated capacity of the unit or device, e.g. an engine, motor, electric cable. (2) To load cargo on a ship to the extent that the load lines are submerged.

overload protection A device which will interrupt the flow of current if it reaches an excessive value.

override A manual means of changing the operation of an automatically controlled device or system.

overshoot The amount by which the maximum instantaneous value of the step function response exceeds the steady-state value.

overshot (1) A fishing tool which is used in a well to pass over a fish (lost tool), grip it, and enable its removal. (2) A joining piece for two lengths of casing.

overspeed trip A device, usually mechanical, which will stop a rotating engine when its speed is about 15% above the rated value. It is also used on some electric motors.

overvoltage A *voltage* above the rated or normal operating value of a device or circuit.

Owner's Declaration A declaration by an owner that he is qualified to own a share in a ship and that no unqualified person owns a share. This declaration must be made when registering the ownership of a British vessel.

oxidation A chemical reaction in which electrons are lost from an atom or oxygen is added to a compound. In a general sense it refers to the corrosion process where metals react with their environment to create, in the case of steel, *rust.*

oxy-acetylene welding The use of a gas flame produced by burning oxygen and acetylene to create a welded joint. A filler rod is used to provide the metal for the joint.

oxygen A colourless, odourless gas which occupies about 20% of the atmosphere. It is essential for the support of combustion and all forms of life.

oxygen analyser A measuring instrument which is used to determine the quantity of oxygen present in an atmosphere or a gas sample. It may be used to ensure that an enclosed space is safe to enter or to establish that a flue gas is inert, i.e. contains less than 5% oxygen.

P

package boiler An auxiliary *fire tube boiler* which is supplied on a base together with feed pumps and other operating equipment in order to minimize installation requirements. See Figure P.1.

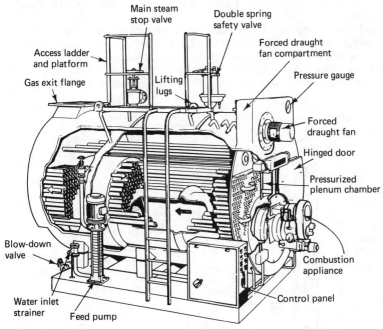

Figure P.1 Package boiler

packer A sealing device which is used to isolate one part of a *borehole* or casing from another to enable testing, improve production or for some other purpose.

packing A material used in *stuffing boxes* to seal against leakage on, for example, a pump spindle. It may be impregrated with a lubricant such as graphite, or made of asbestos or even metal.

paint A protective coating applied to metal surfaces. It consists of a pigment, a binding agent or vehicle and a solvent. The pigment chosen will determine the properties of the paint. The binding agent will determine consistency and ease of application and the solvent makes the paint flow easily.

pallet A board upon which cargo is loaded. It is constructed so that the forks of a fork lift truck can be used to lift and transport it.

Panama Canal Tonnage A *tonnage* measuring system used by the Panama Canal Authorities as the basis for dues paid by a vessel.

panama fairlead An almost elliptical opening formed in a casting which is fitted into a suitably stiffened aperture in the *bulwark*. It is used to bring aboard the mooring lines when in the Panama Canal locks. See Figure B.3.

panamax A vessel of the maximum dimensions to transit the Panama Canal.

panel line A number of specialist work stations which are used for the sequential construction of flat panels which are part of a ship's structure.

panting The in and out movement of a ship's plating.

panting beam An athwartships structural element which is fitted at alternate frame spaces and bracketed to the *panting stringers*. See Figure P.2.

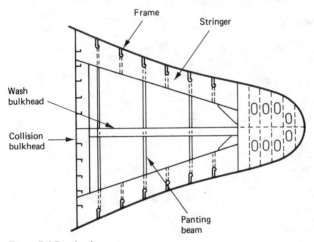

Figure P.2 Panting beam

panting plates Vertical plates which support a steam turbine at one end. They are able to flex or move axially as expansion takes place.

panting stringers Stiffeners which are fitted at about two metre intervals below the lowest deck in the forward region to strengthen the ship's side plating against *panting*. See Figure P.2.

parallel connection (1) The arrangement of cells in a *battery* where all positive terminals are connected together and all negative terminals are connected together. (2) The arrangement of equipment in a system such that one or more may perform a particular function at the same time, e.g. heat exchangers.

140

parallel middle body The ship's length for which the midship section is constant in area and shape.

parallel operation The joining of two or more power sources so that the total output is available to a common load, e.g. alternators.

paramagnetic The ability to become magnetized in the presence of a magnetic field. A material which has a relative *permeability* just greater than one.

parameter (1) A variable that may be given a constant value for a particular purpose or process. (2) One of several variables which are used to measure or describe a process or system.

participation crude The proportion of production allotted to a state in a production sharing agreement.

particular average A partial loss, usually of cargo, which is covered by insurance and is not a *general average* loss.

pascal The derived *SI unit* of pressure which is equal to one *newton* per square metre.

passenger A person on board a vessel at sea who is neither the Master, a member of the crew nor engaged in any way in the business of the ship. A child under one year of age is not considered to be a passenger.

Passenger Certificate A certificate issued to a ship which complies with the safety regulations related to the carriage of passengers.

passenger ship A ship which carries more than twelve passengers.

pay zone A region of *reservoir rock* which contains oil and gas in a concentration sufficient for a commercial exploitation.

payload The cargo carrying capacity of a *container.*

peak pressure indicator An instrument which is used to measure the maximum compression or ignition pressures developed in high speed engines or the peak pressure in a hydraulic system.

peak tank A *tank* located aft of the aft peak bulkhead or forward of the collision bulkhead. It is used for fresh water or sea water ballast.

pedestal A supporting structure for an item of machinery, e.g. a bearing.

penetrameter An image quality indicator which is used when taking radiographs of welds.

Perils of the Sea The hazards of a voyage which are particular to the sea. It is a statement used with reference to the carriage of goods.

period of encounter The wave period relative to the ship when it is moving through the water.

period of roll The time taken during a complete cycle of the *rolling* motion of a vessel.

period of wave The time interval between two wave crests passing the same point.

periodical survey A *survey* undertaken at regular intervals, e.g. annual survey.

peripheral A device connected to a *computer.*

permeability (1) For a compartment on a ship this is the ratio of water which can enter compared with the volume of the empty compartment.

(2) The ratio of the magnetic flux density to the magnetic field strength at a point in a material.

permissible length The length between bulkheads on a ship in order to ensure that it will remain afloat if one, or more, compartments are flooded. The permissible length is some fraction of the *floodable length*. The fraction is called the *factor of subdivision*.

perpendicular See *forward perpendicular* and *after perpendicular*.

petroleum A mineral oil which consists of many different hydrocarbons. It is the raw material from which fuel oils, lubricating oils, paraffin wax and bitumen are obtained. It may also refer to the refined product which is used to fuel motor vehicles.

pH The hydrogen ion concentration of a solution expressed in a manner which gives a scale value of 7 for pure water. Values above 7 are alkaline and less than 7 are acidic.

phase angle The angle between the *current* and *voltage* in an a.c. circuit. It is a lagging value where the circuit contains resistance and reactance and is a leading value when the circuit contains resistance and capacitance. If the phase angle is zero, the current and voltage are in phase.

phase sequence indicator An instrument which indicates the order in which the conductors in a polyphase system reach their maximum current or voltage.

phasor A quantity which varies sinusoidally, when represented as a complex number.

phenolphthalein A reagent used in boiler water testing. It gives a red colour in alkaline solutions.

phosphor-bronze A *bronze* containing about 10–14% tin and a small quantity of phosphorus.It is used in castings when resistance to corrosion or wear is required, e.g. gears, bearings, filter screens.

photoelectric cell A light sensitive sensor which produces an electrical output. Different types of cell exist, namely photoemissive, photo conductive and photovoltaic. They are used in equipment such as the *oil-in-water monitor* and the *crankcase oil mist detector*.

pier See *jetty*.

piezoelectric pressure transducer A sensor which responds to an applied force and produces an electrical output. It is used for dynamic pressure measurement, e.g. combustion pressure in an engine cylinder.

pig A device which is passed through a pipeline. It may clean, clear or inspect the inside of the pipe, signal the passage of a fluid or separate various fluids.

Pilgrim nut A patented form of propeller fastening using a nut which is, in effect, a threaded hydraulic jack. It is used for forcing the propeller onto the tailshaft and can also be used for propeller removal. See Figure P.3.

pilgrim wire A wire which is stretched from the engine to the sterntube and suitably weighted to maintain a tension. It is used to align shafting during installation.

pillar A supporting member between the decks of a ship.

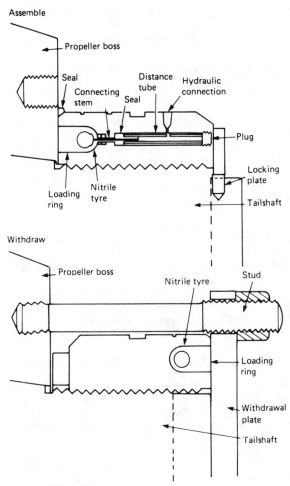

Figure P.3 Pilgrim nut

pilot The statutory definition is 'any person not belonging to a ship who has the conduct thereof.' A pilot is a qualified person who is licensed by the port or National Authority to use his knowledge in assisting a Master to bring his vessel in and out of a port.

pilot ladder The rope ladder arrangement used by pilots when boarding or leaving a vessel.

pilot lamp A lamp which indicates the condition of a particular circuit or item of equipment.

143

pilot valve A small *valve* whose operation will bring about the operation of a larger valve. It is often used in hydraulic circuits.

pilotage The payment made for the services of a *pilot* or a reference to the actual act of using a pilot.

pinion A small gear wheel which engages with a much larger one.

pintle The hinge pin on which certain types of *rudder* swing. See Figure V.1.

pipe-laying vessel A vessel designed and constructed to lay underwater pipelines.

piracy Any violent act or robbery which occurs at sea by the crew of one vessel against another.

piston A cylindrical metal item which reciprocates within a cylinder. It may move as a result of fluid pressure as in an engine, or to compress a fluid, as in a pump or compressor. Leakage between the piston and cylinder is prevented by piston rings. See Figure T.7.

piston crown The upper part of a *piston* which is exposed to the hot gases in an engine.

piston ring A rectangular cross-section ring of cast iron which is cut to enable fitting over a *piston*. The ring fits into a groove and creates a gas tight fit of the piston when moving in the *cylinder*.

piston skirt A thin cylinder of material fitted at the bottom end of a *piston*. It serves to close the ports in a two-stroke engine.

pitch The axial distance travelled by a helicoidal surface during one complete revolution.

pitch ratio The ratio of the pitch of a *propeller* to the diameter.

pitching The rotational motion of a ship about a transverse axis. See Figure R.6.

pitometer A type of submerged *log* which measures ship speed using a *pitot tube*.

pitot tube A tube which is inserted parallel to a flow stream. One orifice faces the flow to receive total pressure and the other registers the static pressure. It is connected to a differential pressure measuring device.

pitting The result of a corrosive or mechanical action which creates small indentations in a metal surface.

planetary gear A system of *epicyclic gears* in which the annulus is fixed, the sun wheel and planet carrier rotate, and the planet wheels rotate about their own axis. See Figure E.3.

planimeter An instrument which mechanically measures the area within an irregular plane figure.

planned maintenance Organizing and scheduling maintenance work in order to minimize lost production and to effectively utilize available resources.

plans Drawings or diagrams which are two dimensional representations of a piece of equipment, a system or an installation.

plant The installation in which a process is carried out. It is often a reference to a collection of equipment.

plasma coating A method of depositing a material such as a metal or ceramic, onto a surface using a gun or torch and a high temperature stream of the substance.

plastic An organic material which can be moulded to shape under the action of heat or heat and pressure. It may be *thermoplastic* or *thermosetting*. Most plastics are resistant to corrosion, have good thermal and electrical resistance but are unsuitable for high temperature use.

plasticity The ability of a material to deform permanently when a load is applied.

plate type gauge glass A high pressure boiler *gauge glass* which uses an assembly of glass plates within a metal housing. See Figure P.4.

plate type heat exchanger A number of ribbed plates sealed together in a frame . The two liquids travel down passages on opposite sides and heat transfer takes place across the plate material.

plates (1) The *electrodes* in a *cell* of a *battery*. (2) Sheets of rolled steel, e.g. ship side plating.

platform See *drilling platform*.

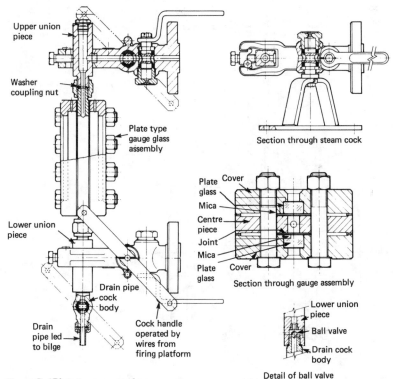

Figure P.4 Plate type gauge glass

145

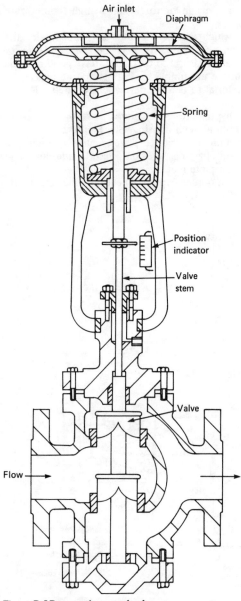

Figure P.5 Pneumatic control valve

146

plating The outer sheet metal covering of the hull of a ship, e.g. bottom plating.

Pleuger rudder An active *rudder* in which a small motor driven propeller is incorporated in a streamlined casing. Ship steering at very low speeds is thus possible and the rudder angle can be greater than 35 degrees.

Plimsoll mark Another term for *load line*. Samuel Plimsoll was the Member of Parliament who introduced the Act of Parliament that made load lines compulsory on British vessels.

plugging The sealing of a *well* which has been abandoned either temporarily or permanently.

plummer block The aftermost propeller *shaft bearing* which has a top and bottom bearing shell fitted. It supports the rotating tailshaft and also some of the propeller weight.

plunger A solid cylinder of metal which slides within a cylinder and creates a fluid pressure. The plunger must be a close fit within the cylinder to create the pressure and minimise leakage.

pneumatic control valve A *valve* which regulates the flow of a fluid. It is remotely operated as the correcting unit of an automatic control system. The actuator is operated by compressed air. See Figure P.5.

pneumatics The use of a gas to transmit a control signal within a system.

pneumercator A remote reading liquid level measuring device using compressed air as the transmission medium. A *manometer* gives a display which can indicate a tank's contents when read against an appropriate scale.

polarity (1) In an electric circuit this term describes which conductor is anodic (positive) or cathodic (negative). (2) When describing parts of a magnet this describes properties existing in certain regions, e.g. north seeking at one end and south seeking at the other.

polarization (1) A change in the electrical state of an insulating material, as a result of an electric field, when each elemental part becomes an electrical dipole. (2) The loss of efficiency in a primary *cell* due to the release of hydrogen gas.

pole (1) A terminal of a *battery*, e.g. positive pole or negative pole. (2) In a rotating machine this is part of the magnetic circuit which usually has an excitation winding.

polychloroprene A synthetic rubber which is used as a sheath for electric cables. It has good mechanical, flame and oil resisting properties.

polymer A material composed of a series of smaller molecular units which may be relatively simple or complex.

polytetrafluorethylene An important fluoropolymer which is chemically inert and heat resistant. It has a low coefficient of friction and is widely used as a bearing material. It can be used dry and is employed in sealed bearings. It may be known as PTFE or by the trade name TEFLON.

polythene A *thermoplastic* polymer of ethylene which is translucent and tough and has good electrical insulating properties. It is often used as weatherproof sheeting.

polyvinylchloride   A *thermoplastic* polymer of vinyl and chloride which is widely used as a plastic, e.g. PVC piping. It has a good resistance to water, acids and alkalis and is also used as an electric insulator.

poop   A *superstructure* at the after end of a ship, usually short in length. The deck may be called the poop or poop deck. See Figure A.1.

poppets   Timber supports for the launching cradle of a ship on an inclined slipway.

port   (1) The left-hand side of a ship when facing forward. (2) A physical communication point between the central processing unit of a *computer* and a *peripheral.* (3) An opening through which fluid enters or leaves the cylinder of an engine, pump or other piece of machinery. It is usually controlled by some form of *valve.* (4) A *harbour* in which ships can load or discharge cargo.

port charges   The dues paid by a vessel for the use of the *port.*

Port of Entry   The *port* at which a vessel or its cargo are cleared by customs for entry into the country.

Port of Refuge   A *port* to which a vessel sails in order to seek a safe place.

Port of Registry   The place where a ship is registered. It is shown on the stern of the vessel.

porthole   A circular opening in the ship's side to provide light and ventilation, and a means of escape. A hinged metal cover or *deadlight* can be clamped over to secure against heavy weather. The term *side scuttle* is also used. See Figure P.6.

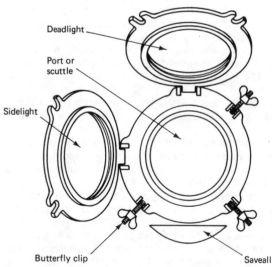

Figure P.6 A porthole

position balance The balancing of linkages and lever movements, usually in a pneumatic device, to achieve an equilibrium condition.

position control The use of a system to achieve control of position or displacement, in a linear or angular sense.

potential difference The difference in electrical state between two points. It is measured by the work done in the transfer of a unit charge from one point to the other.

potentiometer An instrument which may be used to measure or adjust *potential difference* in a circuit.

pounding See *slamming*.

pour point The lowest temperature at which a liquid fuel can be easily handled as a liquid. It is just above the lowest temperature at which the liquid flows under its own weight.

power The rate of conversion, transfer or dissipation of energy. The unit is the *watt*.

power factor In a single phase electrical system this is ratio of active power to apparent power. In an a.c. circuit it is cos ϕ, where ϕ is the *phase angle* between the vectors. In a balanced three-phase system, a motor, or a generator, it is the ratio of the total active power, kW, to total apparent power, kVA.

power factor meter An instrument for measuring the *power factor* in an a.c. electric circuit. It may be calibrated in degrees of phase displacement or values of cos ϕ.

pratique A permit indicating a vessel has complied with medical and quarantine requirements and may, therefore, communicate with the land.

pre-ignition The ignition of the fuel in a petrol engine cylinder by a hot spot before the spark plug has operated.

precision For a measuring instrument this is the extent to which the same input provides the same reading over a number of occasions.

preferential tripping The use of automatic switches to trip or disconnect non-essential loads from the *switchboard* in the event of overload.

pressure The force exerted by a liquid or gas in contact with a surface or boundary, expressed in terms of a unit of area. The *SI unit* of pressure is the *pascal* (Pa) and is a newton per square metre. The *bar* (100 kPa) is also used as a measuring unit.

pressure charging See *supercharging*.

pressure compounding The use of a number of stages of nozzle and blade to progressively reduce the steam pressure in an *impulse turbine*. This results in more acceptable steam flow speeds and an improved turbine efficiency.

pressure maintenance Any means used to sustain the natural pressure of an oil producing well, e.g. water drive or gas drive.

pressure testing The testing of *tanks, watertight bulkheads,* etc., by either water or air pressure, usually to satisfy *classification* requirements.

pressure/vacuum valve A *valve* fitted to the vapour pipeline from a cargo tank. In the event of overpressure it will permit safe venting of vapour. In the event of a partial vacuum it will enable air to be drawn into the tank.

149

primary recovery Oil extraction using the natural forces to drive the oil or gas from the *well*.

primary refrigerant The *refrigerant* which is compressed and expanded in the refrigeration system.

primary winding The electrical winding on the energy input side. It usually refers to the input winding of a *transformer*.

prime mover An *engine* or other device which converts a natural source of energy into mechanical power.

primer (1) The initial *paint* coating which is applied to a bare metal surface. (2) A device which is used to prime, i.e. fill with liquid, a *pump* prior to starting. After priming the pump will continue unassisted as long as a continuous supply of liquid is available. See *water ring primer*.

priming The carry-over of water with the steam leaving a boiler drum. This condition occurs after *foaming*.

principal dimensions The dimensions by which the size of a ship is measured. See Figure P.7.

printer A device which prints characters onto paper.

prismatic coefficient A ship *form coefficient* which is the ratio of the underwater volume to the product of the midship area and the length.

probe An instrument which is used to internally examine an item of equipment such as a condenser tube. See *borescope*.

process The act of physically or chemically changing (including combining) matter or of converting energy.

process control system A *control system*, the purpose of which is to control some physical quantity or condition of a *process*.

process train The treatment process for oil from a production platform prior to its transfer ashore or to a loading facility.

producing horizon The rock which contains oil or gas in exploitable quantities.

production platform A *platform* which receives oil or gas and treats it prior to transportation.

productivity The efficiency of an industrial process, often considered with respect to the workforce.

productivity index A measure of the efficiency of the drive mechanism of a well. It is the ratio of the oil produced in *barrels* per day divided by the fall in pressure in pounds per square inch occurring when a shut in well is made to flow.

products tanker A vessel which has been designed and constructed for the carriage of refined petroleum products. Specially coated tanks are used to enable thorough cleaning and individual tank pumps may be used.

professional engineer The definition adopted by the Engineering Societies of Western Europe and the USA (EUSEC) states:

'A professional engineer is competent by virtue of his fundamental education and training to apply the scientific method and outlook to the analysis and solution of engineering problems. He is able to assume

150

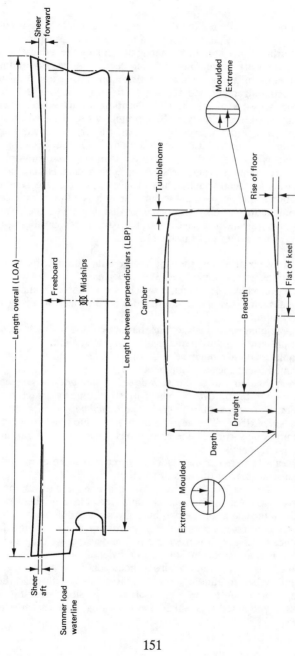

Figure P.7 Principal dimensions

151

personal responsibility for the development and application of engineering science and knowledge, notably in research, designing, construction, manufacturing superintending, managing and in the education of the engineer. His work is predominantly intellectual and varied, and not of a routine mental or physical character. It requires the exercise of original thought and judgement and the ability to supervise the technical and administrative work of others. His education will have been such as to make him capable of closely and continuously following progress in his branch of engineering science by consulting newly published work on a world-wide basis, assimilating such information and applying it independently. He is thus placed in a position to make contributions to the development of engineering science or its applications. His education and training will have been such that he will have acquired a broad and general appreciation of the engineering sciences as well as a thorough insight into the special features of his own branch. In due time he will be able to give authoritative technical advice, and to assume responsibility for the direction of important tasks in his branch.'

profile A drawing of a ship showing the general outline, *sheer* of decks, deck positions and all waterlines. See *lines plan*.

profile cutting machine A machine which is used to cut complex shapes from metal plate. It is usually automatically controlled by paper tape or numerical inputs.

program A related set of instructions that directs and instructs a *computer* in accomplishing specific operations to solve a problem.

programmable read-only memory A non-volatile memory which can be programmed using special equipment.

programme evaluation and review technique A *network analysis* technique which concentrates on events or activities and time estimates associated with them. It is very similar to *critical path method*.

progressive speed trial The testing of a new ship and its engines by determining ship speed, engine revolutions per minute and output power for a range of speeds.

projected area The area of the blades of a *propeller* when projected onto a plane perpendicular to the axis of rotation.

proof stress A value obtained by drawing a line parallel to the elastic stress-strain line for a material, at some percentage of the strain, e.g. 0.1%, the intersection of the two lines gives the proof stress.

propane A colourless gas obtained from *petroleum*. It is easily liquefied and stored under pressure. It is used as a fuel and also in welding and cutting processes. It is an important source of ethylene.

propeller A boss with several blades of helicoidal form attached to it, see Figure P.8. When rotated it 'screws' or thrusts its way through the water by giving momentum to the column of water passing through it. A solid fixed pitch propeller is a single piece casting. See *controllable pitch propeller*.

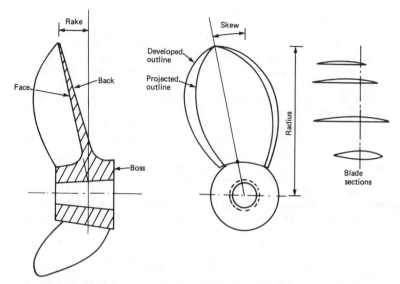

Figure P.8 A propeller

propeller law A series of relationships for an installed power transmission system which refer to *shaft power,* engine speed and engine *mean effective pressure,* i.e. shaft power × (ship speed) ^{3}, shaft power × (engine rev/min) ^{3} and mean effective pressure × (engine rev/min) ^{2}. These relationships place practical limits on an engine designer. An engine under test and coupled to a *dynamometer* will operate according to the propeller law and provide data of use to the ship.

propeller shaft See *tailshaft.*

proportional action The action of a control element which provides an output signal that is proportional to its input signal.

proportional band The range of values of *deviation* which result in the full operating range of output signal of the controlling unit as a result of proportional action only. It may be expressed as a percentage of the controller's scale range.

proportional controller A *controller* which provides only proportional action.

propulsive coefficient The ratio of the *effective power* to drive a ship at a particular speed, to the *shaft power* of the machinery.

protection and indemnity A form of cover against claims, provided to shipowners by associations or clubs formed for this purpose.

protection device A *relay* which detects an abnormal condition in a circuit and operates to protect the circuit or equipment, e.g. *reverse current protection, reverse power protection.*

153

protest See *Noting Protest.*

psychrometer An instrument for measuring relative humidity which uses two thermometers. One is exposed to the air and the other has its bulb kept moist by a water soaked wick fed from a small bath. The two temperatures when plotted on a psychrometric chart will give a reading of relative humidity.

psychrometric chart A graphical display of the properties of moist air, using axes of *dry bulb temperature* and *absolute humidity,* intersected by curves of constant *wet bulb temperature* and *relative humidity.*

pulse A momentary change of a quantity, usually electrical, from its normal value.

pulse controller A *controller* whose output consists only of raise, lower or zero signals. It only provides an output signal when a movement of the *actuator* is required.

pump A machine which provides energy to fluids, usually to produce a flow under pressure. The unit is often further described by its particular method of operation, e.g. axial flow, centrifugal.

pump room A compartment in an oil tanker in which the cargo discharge *pumps* are located. Oil-tight bulkheads are fitted at each end and it is usually located between the cargo tanks and the machinery space.

punching press *A hydraulically powered machine which will create round or elliptical holes and also notches in metal plates.*

purging A cleaning process which usually refers to supplying air to a boiler *furnace* for several minutes prior to lighting-up.

purifier A *centrifuge* which is arranged to separate water and solid impurities from oil.

push rod A metal rod which imparts the motion of a cam follower to the rocker arm of an engine's valve operating mechanism.

PV diagram A graph of the variation of pressure and volume in an internal combustion engine during the operating cycle.

pyrometer A high temperature measuring *thermometer* in which the sensing device does not come into physical contact with the hot body. *Radiation* from the hot body is measured and the instrument is appropriately calibrated.

pyrotechnic The various types of flares and signals used in an emergency to signal other vessels or aircraft.

Q

quarantine   The isolation of a ship until *pratique* is granted. If an infectious disease exists on board, a vessel may be placed at a quarantine anchorage.

quantization   The approximation process in which an analogue signal is converted into a digital signal.

quarl blocks   *Refractory* bricks of a special shape which are fitted around the *burner* openings of a boiler *furnace*.

quarter deck   A raised upper *deck* at the after end of a ship. It is usually a feature of smaller vessels.

quarters   The accommodation area for the crew of a ship.

quasi propulsive coefficient   The ratio of the *effective power* necessary to drive a ship at a particular speed to the power which must be supplied at the propeller, i.e. the *delivered power*.

quay   An artificial solid construction, alongside or projecting into a *harbour* or basin, which acts as a landing place for cargo and passengers. It may also act as a fitting out or repair place for a ship which is moored alongside.

quick closing valve   A *valve* which has a collapsible *bridge* and can therefore be shut quickly from a remote point. A manually operated wire or a hydraulically operated cylinder may be used to close the valve. See Figure Q.1.

quill shaft   An assembly of a hollow shaft rotating on a solid spindle. The two shafts are joined at one end, usually by a *flexible coupling,* and the drive is provided to the solid shaft at the opposite end. It is often used in helical *double reduction gears*. See Figure Q.2.

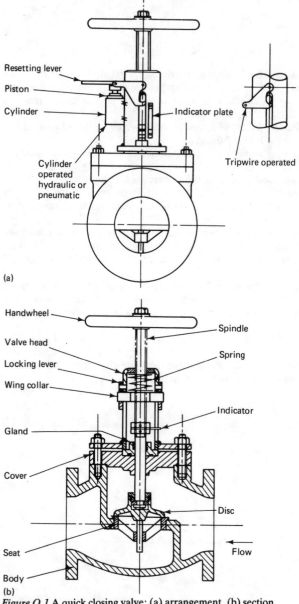

Resetting lever
Piston
Cylinder
Cylinder operated hydraulic or pneumatic
Indicator plate
Tripwire operated

(a)

Handwheel
Valve head
Locking lever
Wing collar
Gland
Cover
Seat
Body
Spindle
Spring
Indicator
Disc
Flow

(b)

Figure Q.1 A quick closing valve: (a) arrangement, (b) section

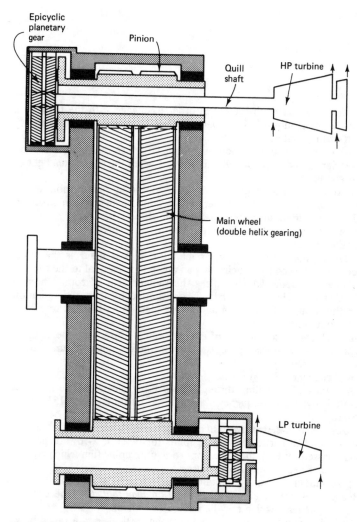

Figure Q.2 A quill shaft

R

rack A bar in which teeth are cut in order to engage with the teeth of a *pinion*. A linear motion of the bar will then be converted into a rotary motion of the pinion.

racking The transverse distortion of a ship's structure due to acceleration and deceleration during *rolling*. It is greatest when the ship is in a light or *ballast* condition.

radar RAdio Detection And Ranging. A system using pulsed radio waves which, when reflected or regenerated, produce a display on a screen which can provide a distance and direction of the target.

radial flow A flow of fluid from the centre of a circular body to the periphery or vice versa.

radian The *SI unit* of plane angular measure. It is defined as the angle subtended at the centre of a circle by an arc of length equal to the radius.

radiant boiler A *boiler* which utilizes the radiant heat of combustion to produce steam. This radiant heat is transmitted by infra-red radiation within the furnace. Roof firing and a very high boiler is usual to ensure efficient operation.

radiation (1) The dissemination of energy from a source, e.g. heat, light, X-rays. (2) A heat transfer process where the intervening medium, usually air, is not heated.

radiation pyrometer See *pyrometer*.

radio A means of signalling through space using electromagnetic waves generated by high frequency, alternating currents, i.e. 15 kHz to 100 MHz.

radiography A *non-destructive testing* process which examines materials, structures and in particular welded joints for defects. *X-rays* or gamma rays are passed through the object onto a photographic film which creates a record of the test. See *penetrameter*.

rail (1) A pipeline, usually containing fuel, which is maintained at a particular pressure or acts as a distribution *manifold*. (2) The capping applied to *bulwarks* which may be of wood or metal.

rake An inclination from the horizontal or the vertical, e.g. a raked bow.

ram (1) A hydraulically operated piston which seals off a well when the *blowout preventer* is actuated. (2) The piston rod of a hydraulic cylinder as in a ram-type steering gear.

ramp A hinged platform which enables wheeled vehicles to drive from the quay into the hold of a *roll-on roll-off ship*. It is hydraulically operated and can be adjusted to suit changes in the tide during loading.

random access memory A volatile electronic *memory* that can be erased and modified.

range The difference between the upper and lower limits of an instrument's displayed measurements.

ranging Adjusting an instrument so that the index movement is in agreement with a scale at two or more positions.

Rankine cycle The ideal practical cycle for steam turbines and steam engines. It assumes all operations are carried out in their separate units, e.g. steam is condensed in the condenser at constant pressure, feed water is supplied by the feed pump at constant volume and appropriate pressure rise, etc.

Rapson's slide A *crosshead* arrangement used on ram-type steering gear in which the mechanical advantage increases with the angle of turn. It is used to convert the straight line motion of the rams into an angular movement of the tiller.

rate action See *derivative action*.

Rateau turbine An *impulse turbine* with several stages, each stage being a row of nozzles and a row of blades, i.e. pressure compounded.

rated output According to British Standard 649 : 1958, oil engine types, this is the brake power output that can be maintained for 12 hours at rated speed. The stated conditions are a temperature of 29.4°C, an atmospheric pressure of 749 mm of mercury and a humidity of 15 mm of mercury vapour pressure.

rating (1) A specified limit with respect to operating conditions. It may relate to a device, a system, an electrical or mechanical machine, etc. Propulsion engines have various levels of rating, e.g. normal *continuous sea-service, maximum continuous service,* and *trial trip* or short-time rating. (2) Any crew member who is not an officer.

ratio control system A control system in which two or more physical quantities or conditions are maintained at a predetermined ratio.

reactance The component of *impedance* in an a.c. circuit which is due to *inductance* and/or *capacitance*.

reaction turbine A *turbine* in which the steam is expanded as it passes between rows of fixed and moving blades. The narrowing space between the blades acts as a nozzle, creating a reaction force in addition to the impulse force. It is more precisely referred to as an impulse-reaction turbine.

read To obtain information from an *input device* or the *memory* of a *computer*.

read only memory A non-volatile computer *memory* whose contents are fixed during manufacture.

read out The removal of data from a computer *memory*.

reciprocate An alternating motion between two extreme points on a fixed path, e.g. up and down.

reciprocating pump A positive displacement *pump* in which a piston moves in a cylinder alternately drawing in liquid and then pumping it out. Suction and discharge valves are operated by the liquid pressure. The pump may be single or *double-acting*.

159

recirculating feed line A pipeline which enables the recirculation of feed water back to the condenser. It is used during reduced power operations to provide the cooling necessary for various steam returns and is usually automatically controlled.

recording instrument A measuring instrument which records the values of the measured quantity on a chart.

recovery factor The ratio of the economically recoverable oil or gas in a reservoir to the oil in place.

rectifier A device which converts an alternating current into a unidirectional or direct current. This may be done by suppressing alternate half-waves or by inverting alternate half-waves. See *bridge rectifier, full-wave rectifier, half-wave rectifier*. See Figure R.1.

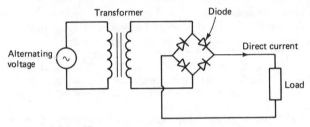

Figure R.1 A rectifier

reducing valve A *valve* which is designed to maintain a constant lower fluid pressure at outlet, regardless of the inlet conditions.

reduction gearing An arrangement of gears which converts a high speed, low torque input into a low speed, high torque output. See *double reduction gearing, epicyclic gears, gear box, gear train, gear wheel, helical gearing, worm gear*.

redundancy The provision of equipment or components which are surplus to operational requirements in order to continue operation after a failure. This will improve *reliability* and also safety.

Redwood seconds A unit of measurement for *kinematic viscosity* when a Redwood viscosimeter is used. Standard temperatures of 70°F (21°C) for lubricating oil and 100°F (37°C) for fuel oils are used. See also *centistokes, viscosimeter*.

reefer A *refrigerated general cargo ship*.

refit A complete maintenance and refurbishing process usually requiring equipment overhaul and repair, cleaning and painting. Improvements and modifications are usually undertaken to modernize or up-date fixtures and fittings in, for example, the refit of a ship.

refractory A material, usually in brick form, which is used to line boiler furnaces. It must resist high temperatures, change in temperature and the effects of hot flowing gases. Suitable materials include china clay, fire clay and silica.

refrigerant A substance suitable for use in a *refrigeration* system. It must boil at a low temperature and reasonable pressure and condense at, or near, normal sea water temperature at a reasonable pressure. It must also be free from toxic, explosive, flammable and corrosive properties where possible. See *freon.*

refrigerant dryer See *dryer (2).*

refrigerated general cargo ship A vessel designed and constructed for the carriage of general or refrigerated cargo. It will be similar in construction to a *general cargo ship,* with perhaps additional tween decks, and all cargo holds will be insulated. The *refrigeration* system will enable the carriage of cargoes at different temperatures.

refrigeration A process in which the temperature of a space or its contents is reduced to below that of their surroundings.

regenerative condenser A *condenser* in which some of the steam bypasses the cooling medium and is used to reheat the condensed steam thus minimizing *undercooling.* See Figure R.2.

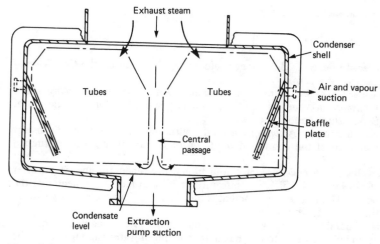

Figure R.2 A regenerative condenser

register (1) A short-term *memory* storage unit in a *computer.* (2) The complete oil burning unit which is secured to the boiler furnace front or roof. See Figure R.3. (3) A book which lists details of almost all ships over 100 tons, i.e. *Lloyd's Register of Shipping.*

registered ship A ship over 15 tons and other than a river, coastal or fishing vessel, must be registered with the Registrar General of Shipping in order to comply with the Merchant Shipping Act.

161

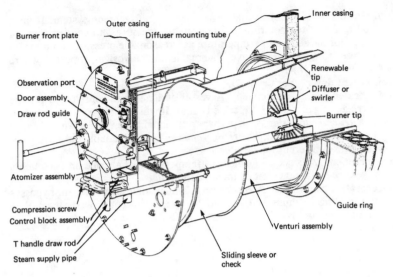

Figure R.3 A register

registered tonnage The *tonnage* value which is stated on the *Certificate of Registry* of a ship.

Registrar General of Shipping and Seamen An official of the *Department of Transport,* Marine Directorate, who is responsible for the register of British ships and their crews.

regulating system A *control system,* the purpose of which is to hold constant the value of the controlled condition or to vary it in a predetermined manner.

reheat boiler A *boiler* used in conjunction with a reheat turbine. Steam after expansion in the high pressure turbine is returned to a reheater in the boiler. The heat energy content is thus raised before the steam passes to the low pressure turbine.

relative humidity The ratio of the amount of water vapour present in a given volume of air to the maximum amount of water vapour that can be present before precipitation occurs.

relay (1) An electric switch which, when operated, brings about a change in an independent circuit. (2) A pneumatic amplifier.

release note An authority which permits the Master of a vessel to deliver the cargo.

reliability The ability of a component, a machine or a system, to perform a particular function under stated conditions for a particular period of time. It may be given as a probability or a success ratio.

relief valve A *valve* which will open automatically to release excess pressure from a pipeline or containment vessel, e.g. safety valve on a boiler drum.

162

relief well A directional well which is drilled into a well that has suffered a blowout. Heavy drilling fluid can then be pumped in to kill the blowout well.

reliquefaction The liquefying by refrigeration of cargo 'boil-off' from a *liquefied gas carrier*.

reluctance The ratio of the magnetomotive force to the magnetic flux around a magnetic circuit.

repeatability The ability of an instrument to reproduce readings during a short duration test under fixed conditions.

repressuring The injection of fluid into a hydrocarbon reservoir in order to increase the pressure.

reserve buoyancy Enclosed spaces which provide *buoyancy* in addition to that required by a vessel to float. It is a consideration in the assignment of *freeboard* to a ship.

reserves The amount of crude oil or gas which can be profitably recovered from a reservoir.

reservoir rock Any permeable or porous rock which has oil or gas that can be extracted from its pore spaces. Examples are sandstone and limestone.

reset action See *integral action*.

residual carbon A fuel quality usually determined by the *Conradson carbon* residue test. The residual carbon is a measure of the fuel's burnability.

residual fuel A heavy fuel which is the residue of the crude oil refining process. It is a poor quality fuel containing impurities introduced during the refining process. It is extremely viscous with a viscosity in the region of 3500 seconds Redwood No. 1 at 37°C, or higher. Special fuel treatment is necessary prior to combustion.

residual magnetism The magnetism retained by certain materials after the magnetizing force has been removed.

residuary resistance The difference between the total and the frictional resistance of a ship. It is largely made up of wave-making resistance. See *eddy-making resistance, frictional resistance, wavemaking resistance.*

residue (1) The product remaining after the various refining processes have been completed. (2) The waste or remainder following a process, e.g. tank cleaning.

resilient mountings Machine supports which utilize springs or an elastomeric material to prevent the transmission of vibration to the supporting structure.

resin A hard, brittle substance which is insoluble in water. In precise terms a resin is added to a polymer prior to curing. See *epoxy resin.*

resistance (1) The property of a substance which restricts the flow of electricity in a circuit and dissipates energy. It is measured in *ohms*. (2) Any form of opposition to a fluid flow, e.g. *frictional resistance.*

resistance thermometer A temperature measuring instrument using metals such as platinum or nickel in an electric circuit. An increase in temperature will cause an increase in resistance and a *Wheatstone bridge* will measure the resistance change and indicate the temperature. The

163

measuring range is from −200 to 600°C.

resistor A device which is introduced into an electric circuit to produce resistance.

resolution (1) The smallest change in input that can be detected by a measuring instrument. (2) The quality of a visual image produced for example on a radar screen or computer monitor.

resonance A condition associated with *vibration* or oscillation where the forcing impulses occur at the *natural frequency* of the component or structure. If significant *damping* is not present the amplitude of the vibration or oscillation will increase continuously. Resonance may also occur in acoustic systems and electrical systems.

resonant frequency (1) The frequency, in a control system, at which the ratio of the amplitude of the *controlled condition* to the *command signal* is a maximum. (2) The *frequency* which causes resonance in a particular system.

Respondentia The use of a ship's cargo as security for a loan needed by the ship to complete the voyage. A Respondentia Bond is issued, which is a lien on the cargo. It is void should the vessel be lost.

Restraint of Princes, Rulers and Peoples A reference in the *Hague Rules* to any action by Monarchies, Governments, Dictatorships, etc., against a vessel, e.g. arrest.

reverse current protection A *relay* fitted between a d.c. generator and the switchboard which will trip and disconnect the main circuit breaker in the event of a reverse current flow.

reverse osmosis The use of a high pressure pump to force a liquid through a semi-permeable membrane, which will not pass salts or dissolved solids. It is a means of producing distilled water from sea water. See *osmosis*.

reverse power protection A *relay* which monitors the direction of power flow from a generator to the switchboard. In the event of a reverse power flow it will trip and disconnect the main circuit breaker.

Reynolds number A dimensionless index used to describe the nature of liquid flow. It is the product of the average velocity, density and internal diameter of the pipe divided by the fluid viscosity.

rheostat A *resistor* whose resistance can be varied whilst connected in a circuit.

rig See *drilling rig*.

righting lever The length of the lever arm between a vertical extended up from the *centre of gravity* to the vertical extended up from the *centre of buoyancy* of a ship. It is referred to as GZ and is the criterion used to determine *stability* at large angles of inclination.

ring main A complete ring of piping with branches off to various tanks and also to discharge and loading lines. The use of several ring mains enables flexibility in the loading and discharge of various 'parcels' of oil.

ring press See *gap press*.

rise of floor The height of the bottom shell plating above the *base line*. It is measured at the moulded beam line of a ship. See Figure P.7.

riser tensioner A support system for the *marine riser* pipe, to maintain it in tension regardless of the heaving of the platform. A system of wires, sheaves and pneumatic cylinders is used for each of the four or six tensioners installed. See Figure R.4.

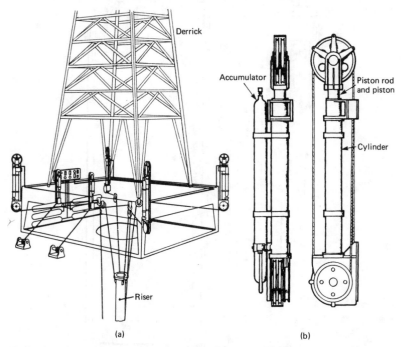

Figure R.4 A riser tensioner: (a) arrangement, (b) construction

rising main A vertical section of *fire main* piping.

rock bit A drilling bit for use in hard rock formations.

rocker arm A metal bar which pivots on the rocker shaft. One end is moved up and down by a push rod riding on a cam follower. The other end operates the inlet or exhaust valve of a four-stroke engine. See Figure T.7.

roll-on roll-off ship A vessel designed and constructed for wheeled cargo, usually in the form of trailers. Large bow or stern ramps enable the cargo to be driven on board. *Containers* are often carried on the deck of this type of vessel. See Figure R.5.

roll stabilization The reduction of the *rolling* motion of a ship by creating an opposite force to that creating the roll. See *stabilizer*.

rolling The rotational motion of a ship about a longitudinal axis, which is usually confined to a few degrees either side of vertical. See Figure R6.

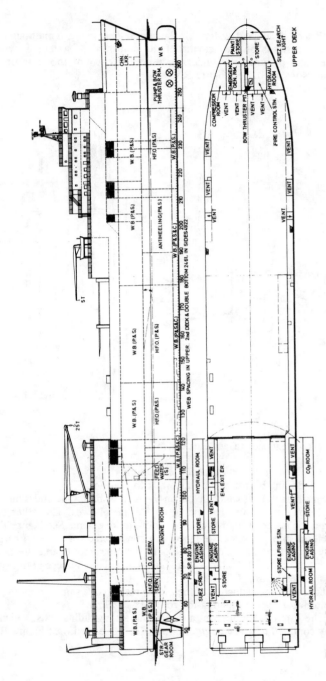

Figure R.5 A roll-on roll-off ship

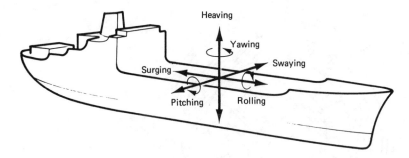

Figure R.6 Ship motions

roof fired boiler A *boiler* in which the burners project into the furnace from the roof.

root-mean-square A measure of variable electrical quantities. It is the square root of the mean value of the squares of all the instantaneous values taken over a complete cycle.

rope guard A circular steel plate made in two halves and fitted around the stern frame boss to enclose the propeller shaft and seals. A very small clearance is provided between the stationary rope guard and the rotating propeller to prevent the entry of mooring ropes or other debris.

rotameter A *variable area flow meter* in which the flowing liquid displaces a float. The specially shaped float rotates and moves up or down in a graduated glass tube, its position giving a flow measurement.

rotary compressor A *compressor* which uses a rotating element, e.g. a cam or a series of vanes, to bring about pressurizing of the gas. Rotary air compressors and refrigerant compressors are in use.

rotary converter A machine which converts a.c. to d.c. It combines motor and generator action in a single armature winding which is connected to slip rings and a commutator. It is excited by a single magnetic field.

rotary pump Any type of *pump* using a rotating element to displace liquid, e.g. vane pump, screw pump.

rotary table An assembly which is mounted in the derrick floor and rotates the *drill string*. It also permits the *kelly* to be raised and lowered through it.

rotary vane steering gear A hydraulic steering mechanism which consists of a vaned rotor that moves within a vaned stator. Chambers are formed between the vanes on the rotor and those on the stator. The rotor will move when oil is supplied into appropriate chambers and removed from others. See Figure R.7.

rotating cup burner A boiler *burner* which atomizes and swirls the fuel by throwing it from the edge of a rotating tapered cup.

Rotocap See *valve rotator*.

rotor The rotating part of a machine, e.g. a steam turbine.

167

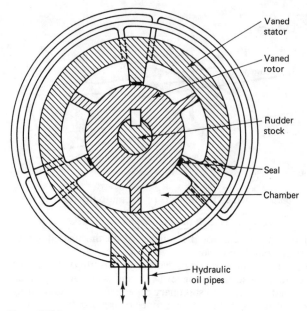

Vaned
stator

Vaned
rotor

Rudder
stock

Seal

Chamber

Hydraulic
oil pipes

Figure R.7 A rotary vane steering gear

roundtrip The complete process of removing (pulling out) and replacing (running back) the *drill string* when, for example, changing the bit.

Royal Institution of Naval Architects A learned society founded in 1860 to promote the improvement of ships and all that specially appertains to them. It is a nominated and authorized body of the *Engineering Council* and appropriately qualified members are entitled to be registered as *Chartered Engineers*. The headquarters are at 10 Upper Belgrave Street, London SW1X 8BQ, UK.

royalty A tax which is paid to the owner of mineral rights, usually a government when offshore rights are involved.

rubber A tree sap which solidifies to form a tough, elastic material which is unaffected by water but is attacked by oils and steam. It is used as a jointing material for fresh and salt water pipes and also in water lubricated bearings.

rubbing strake An extra length of plate fitted in order to make contact first and prevent damage to the main structure. An example would be a length of plating along the side shell of a ship where it may make contact with harbour walls.

rudder A pivoted structure which acts as a control surface to direct a ship's movement in a horizontal plane. There are various types which are described by the arrangement about the turning axis. See **balanced rudder,** **semi-balanced rudder, unbalanced rudder.**

rudder axle   The solid shaft about which a **balanced rudder** rotates.

rudder carrier   The bearing which supports the rudder weight. The upper part is keyed to the **rudder stock.** The lower half is bolted into an insert plate in the deck of the steering flat and is chocked against fore and aft and athwartships movement. See Figure R.8.

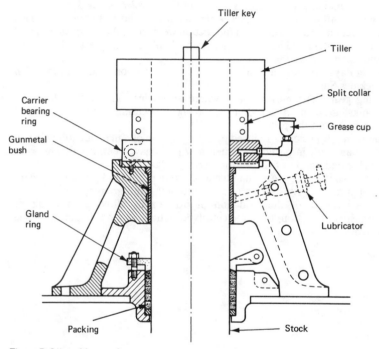

Figure R.8 A rudder carrier

rudder coupling   The horizontal surface at the top of the rudder to which the palm of the **rudder stock** is bolted.

rudder horn   That part of the **stern frame** which supports a **semi-balanced rudder.**

rudder post   The vertical part of a **stern frame** to which the rudder is attached. It is sometimes called the **stern post.**

rudder stock   The shaft which connects the **rudder** to the **steering gear.** The lower end is opened out into a palm for coupling to the rudder and the upper end is keyed to the **tiller.**

169

rudder stop A lug, bracket or other mechanical device which will limit the rudder movement to about 37 degrees port or starboard.

rudder trunk An open region in the stern for the entry of the *rudder stock* into the steering flat. A horizontal platform may be fitted midway up the trunk to fit a watertight gland. The trunking above is then constructed to be watertight.

Rules of Practice The rules used for the adjustment of *particular* and *general average*.

run The immersed body of a ship, aft of the *parallel middle body*.

run on, run off plates Extra pieces of metal positioned at the start and finish of a weld in order to ensure a full section weld along the complete seam of the main plates. The run on and run off plates may be used to make specimens for testing.

running days A term used in charters to describe consecutive days which includes Sundays.

running gear The moving parts within an engine. This will include the crankshaft, connecting rod, main and big end bearings, any crosshead and bearings, and the piston.

running in A period of reduced power operation of a machine following an overhaul and, in particular, when new parts have been fitted. This is to enable bedding in of surfaces to occur gradually.

running maintenance Any maintenance action performed whilst equipment is operating. It is usually a reference to oiling and greasing.

rust A product of the oxidation of iron or its alloys as a result of atmospheric action or an electrolytic action around impurities. See *corrosion*.

S

sacrificial anode A block of aluminium or zinc alloy which is electrically connected to a ship's hull to act as the anode of an electrolytic cell and be eaten away. The steel hull is therefore protected. See *electrochemical corrosion.*

saddle tank A *tank* which is almost triangular in cross-section and extends from the deck into the hold of a bulk or combination carrier on either side of the centreline. It is usually arranged for the carriage of water ballast in order to raise the ship's centre of gravity when carrying a light cargo.

SAE number A classification system for lubricating oils which is based on *viscosity*. It was introduced by the American Society of Automotive Engineers. A thick viscous oil has a high SAE number at normal engine operating temperature.

safe port A port in which a vessel and its cargo may remain safely afloat without interference from any physical or political cause.

safe working load A maximum working load which must not be exceeded. It is usually clearly indicated on lifting equipment.

safe working practice An accepted working procedure which will ensure that a person is able to perform a particular task safely and is not exposed to any unnecessary risks. Various codes of safe working practices are issued by national authorities for specific industrial situations.

safety valve A *valve* which is designed to open automatically and relieve excess pressure. Spring-loaded valves are always used on board ship because of their positive action at any inclination. Boiler safety valves are always fitted in pairs, usually on a single valve chest. See Figure S.1.

safety zone The area around an offshore installation, within a radius of 500 m. Ships are prohibited from entry except under special circumstances, e.g. when in distress.

sagging The condition of a floating ship when the distribution of weight and buoyancy along its length is such that the weight amidships exceeds the buoyancy.

salient pole A field pole which projects from the hub. This type of rotor construction is used with medium and slow shaft speeds and is common on marine generators.

salinity A measure of the amount of salt present in water, usually given by a *salinometer.*

salinometer An instrument which measures and indicates the amount of sodium chloride, expressed in parts per million, in a given sample of water. Electrical *conductivity* is the usual means of measurement.

salt dome A plug of rock salt which has been forced up through other rock strata until it lies beneath cap rock. The salt dome may trap oil or gas and such a formation, when discovered, will often be investigated.

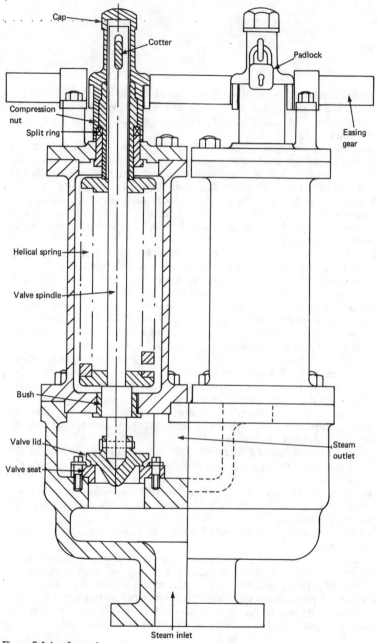

Figure S.1 A safety valve

salvage The money paid for assistance in saving a ship or other goods from danger at sea or the actual goods themselves.

Salvage Association A body which acts, through its Surveyors, to establish the nature, cause and extent of damage to a vessel and advise regarding repair or any means of determining the extent of loss.

sampling connection A water outlet cock and cooling arrangement for drawing off *feed water* samples from a boiler.

samson post A rigid vertical post used in place of a mast to support derricks. It may also be known as a *king post*.

sand blasting An abrasive cleaning method for steel plating which may use dry sand or a sand and water mixture.

sanitary pump A *pump* which supplies sea water for flushing water closets and urinals.

saturated steam *Steam* which is at the boiling temperature associated with its particular pressure, i.e. the saturation temperature. It will contain small quantities of water and is considered to be wet steam.

saturation (1) The condition of air at any given temperature and water vapour pressure, when a reduction in temperature would cause condensation. (2) A limiting condition in many electrical items, e.g. when the output current of an electronic device is constant and independent of the voltage.

saveall Any receptacle positioned to collect leaking liquid from tanks, pipes, machinery, etc.

Saybolt A *viscosity* measurement (US) which may be Saybolt Universal (100°F) or Saybolt Furol (122°F) with the latter being used for drilling fluid and other viscous oils. The unit is seconds.

scale (1) A numerical factor which relates the measured value to the actual value. (2) An oxide coating on the surface of a metal, e.g. mill scale. (3) Deposits of salts and impurities from feed water onto the heat exchange surfaces, e.g. the tubes, of a boiler.

scantlings The dimensions of the structural items of a ship, e.g. frames, girders and plating.

scattered light A technique used in the measurement of the oil content of a water sample. The light is almost infra-red.

scavenge belt The enclosed region around the air inlet ports of a two-stroke, slow speed diesel engine, which extends the length of the engine. It is used to store low pressure air from the turbo-blowers prior to its entry into the engine.

scavenge fire A fire in the scavenge space of an engine. It is usually due to the combustion of excess amounts of cylinder oil, which have collected in the space, being ignited by hot gases from the cylinder blowing past the piston rings.

scavenging The removal of exhaust gases from a two-stroke engine by blowing in fresh air.

Schottel rudder A Z or double right-angled drive mechanism in which the propeller operates within a duct. The propeller and duct can be rotated in

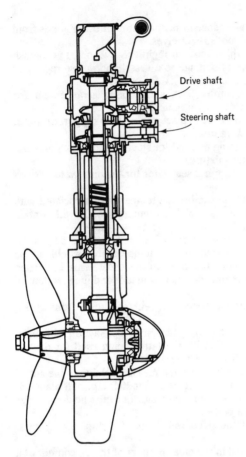

Labels on figure: Drive shaft, Steering shaft

Figure S.2 A Schottel rudder

order to steer the driven vessel. See Figure S.2.

scoop cooling A large opening in the bottom shell which admits sea water into a large volume, low pressure circulating system. The forward motion of the ship will draw in water and a circulating pump is only required when the ship is moving slowly or stopped.

Scotch boiler A large diameter, circular firetube boiler of relatively short length. It is all but obsolete now.

scraper ring A *piston ring* fitted at the lower end of a piston of a trunk piston engine. It is used to remove excess quantities of lubricating oil from the cylinder walls and return it to the crankcase.

screen plate A metal plate of heat resisting steel, through which the superheater tubes of a boiler pass. It shields the headers from the direct heat of the furnace and also minimizes the escape of gases into the machinery space.

screen tubes A number of rows of metal tubes which shield the superheater from the radiant heat of the furnace.

screw A *propeller*.

screw aperture The opening aft of the *stern frame* in which the propeller rotates.

screw compressor A *compressor* which uses the intermeshing of rotors of helical gear-type form in a casing. A single rotor and star wheels may also be used to achieve balance. It is often used for refrigerants but requires a large oil separator on the discharge side.

screw displacement pump A *pump* which uses the intermeshing of rotors of helical gear-type form in a casing. The liquid or air is forced around between the casing and the space between teeth.

screw down non-return valve A *valve* in which the disc is not attached to the spindle. The disc must have wings or a guide to ensure it seats correctly when the valve is closed. The disc will reseat automatically if a reverse flow of liquid occurs. See Figure S.3.

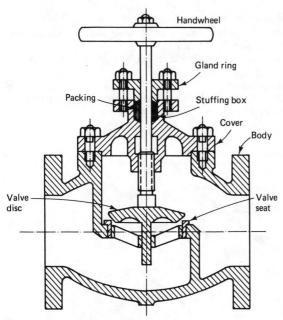

Figure S.3 A screw down non-return valve

175

screw effect   A sideways thrust which results from the *propeller* rotation and affects the steering of a ship. It is most noticeable when near to a quay or in a narrow channel.

screw lift valve   A *valve* in which the disc is attached to the spindle and will be fixed in any position set by moving the spindle.

screw race   The turbulence in the water leaving the propeller.

scrubbing tower   A cleaning chamber in an *inert gas plant*. Sea water is used to scrub or remove contaminants from the gas before it passes to the cargo tanks.

scum valve   A *boiler mounting* which is used to remove scum from the water surface. A shallow dish is positioned at the normal water level and connected to the valve.

scupper   A deck drain to remove sea water, rain water or condensation.

scuttle   (1) An opening in a deck, hatchway or the ship's side to enable access or escape from a space. *Portholes* are also called side scuttles. (2) To deliberately sink a ship by letting in water.

SD14   A particular design of *general cargo ship* which was intended as a replacement for the *Liberty ships* built in the 1940s. It was designed by Austin and Pickersgill (UK) and hundreds were built by them.

sea cock   Any *valve* in the ship's side which permits the entry of sea water.

sea-going   The ships or the men who travel on deep sea voyages across the oceans.

sea inlet box   A steel structure which is fitted into the side shell below the water line and is open to the sea. *Sea tubes* or flanges are fitted onto the box to attach *sea cocks*. See Figure S.4.

sea kindliness   A description of a ship which performs well in various sea conditions and bad weather.

sea spectra   See *wave spectra*.

sea suction   The aperture through which sea water is drawn or the part of a pumping system which will admit sea water into it.

sea trials   See *trial trip*.

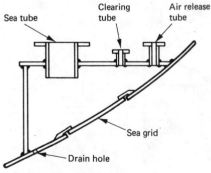

Figure S.4 A sea inlet box

sea tube A thick walled steel tube with a flange on the inboard end for mounting a *sea cock*. The outboard end is welded into a *sea inlet box*. See Figure S.4.

SEABEE A particular design of barge carrying ship. Thirty eight barges may be loaded, each carrying up to 1000 tonnes of cargo. They are loaded by an elevator located aft and winched along the various decks.

seafarer A person who works on ships for a living.

seakeeping The many aspects of a ship's design and construction which determine its ability to operate efficiently at sea, e.g. stability, strength and speed.

seal pot A chamber used with an instrument which measures the flow of a high *viscosity* liquid. The viscous liquid is in the upper half and a low viscosity non-mixing liquid is in the lower half which is connected to the instrument. See Figure S.5.

seam A horizontal weld in shell plating.

seaman Any person serving on a merchant ship except the *master*. It is often used to describe a seafarer who is neither officer nor apprentice.

seat The structural support for an item of machinery or equipment.

seaworthy A vessel which is fit in all respects for the anticipated perils of the voyage and will carry the cargo and crew in safe condition. It also refers to fitness to receive, carry and preserve the particular cargo.

second moment of area A calculation, which for the irregular areas of a ship is done using *Simpson's rule*. It is found by Σar^2 where a is an element of area and r is the perpendicular distance from the axis considered.

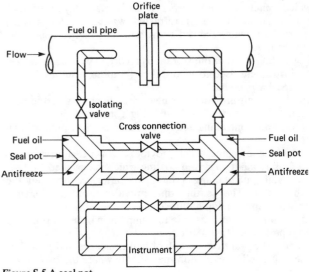

Figure S.5 A seal pot

177

secondary barrier A structure which is capable of retaining the cargo of a *liquefied gas tanker* in the event of main tank fracture. It is a *classification society* requirement for this type of ship.

secondary recovery Any technique used to extend the life of an oil well when the natural pressurized flow has reduced. Water or gas may be injected to re-pressurize the well and enable the recovery of more oil.

secondary refrigerant A liquid or gas which is cooled in the evaporator of a *refrigeration system* and then circulated through the refrigerated space, e.g. *calcium chloride* brine or air.

secondary winding The winding of a *transformer* which delivers energy to the load.

section board A grouping of electrical services which are fed from the main *switchboard*.

section modulus The ratio of the *second moment of area* to the distance measured from the *neutral axis* to the extreme edge of the section, e.g. the deck or bottom plating of a ship.

sedimentary rock Rock which has formed by the accumulation of material from older rocks, deposits from thermal and chemical action and from organic matter. The organic matter in this rock is believed to be the source of hydrocarbon, i.e. source rock. The oil usually migrates from the source rock until it collects in an oil trap.

segregated ballast tanks *Tanks* which can only be used to carry sea water *ballast*. An oil tanker must have a sufficient number of such tanks to be able to achieve a safe operating condition on a ballast voyage.

seismic survey A geophysical survey using an energy source to create explosions which generate shock waves in the surface layers of rock beneath the sea. The reflected or refracted waves are detected by *hydrophones* and the signals received provide details of the depth and extent of *sedimentary rock*.

seismogram A pictorial record of the vibrations recorded by a *seismometer*.

seismometer An instrument used to record the vibrations resulting from a *seismic survey*. See *hydrophone*.

self-acting controller A unit providing a single variable control action with no separate source of power. The fluid being controlled provides the means of achieving the controlling action, e.g. a pressure regulator in an air supply line.

self-excited An electrical machine which provides its own excitation.

self-polishing anti-fouling paint A *paint* coating which is designed to wear down smoothly while maintaining a bio-active interface between the coating and the water. The coating thus provides protection against marine growth whilst minimizing hull resistance.

self-priming pumps A *pump* which is able to pump air in order to prime or fill the suction piping when first started.

self-propulsion point The point on a graph of hull resistance and propeller thrust to a base of propeller revolutions per minute where the two curves meet.

178

self-supporting tank A *tank,* used for the carriage of liquefied gas, which is strong enough by virtue of its construction to accept any loads imposed by the cargo it carries.

self-trimming A cargo hold with a large open hatch to enable loose bulk cargoes to be loaded and directed into all parts of the hold.

selsyn See *synchro.*

semi-balanced rudder A *rudder* design, see Figure S.6, where part of the area is positioned forward of the turning axis, e.g. a spade type rudder. See *balanced rudder.*

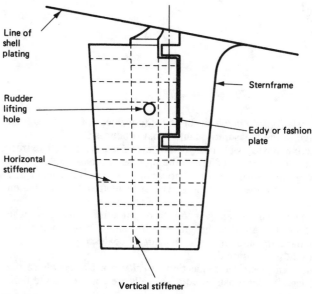

Figure S.6 A semi-balanced rudder

semi-membrane tank A *tank,* used for the carriage of liquefied gas, which requires the insulation between the tank and the hull to be load bearing. It is almost rectangular in cross-section and unsupported at the corners.

semi-submersible drilling unit A floating *drilling rig* which consists of hulls or caissons which carry vertical columns that support the drilling platform and all its equipment. It will operate in deep water when it is ballasted to a draught of about 25 metres and provides a very stable drilling platform. See Figure D.2.

semiconductor A material which is neither *conductor* nor *insulator* but is somewhere in between. It is used in a large number of solid-state devices such as *diodes, transistors* and *integrated circuits.*

179

senhouse slip A quick-release fastening arrangement used, for example, on lifeboat *gripes*. Slipping a ring will release a metal clip and the link of chain it is holding. See Figure S.7.

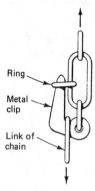

Ring

Metal clip

Link of chain

Figure S.7 A senhouse slip

sensitivity The relationship between the index movement of an instrument and the change in the measured quantity that produces it.

sensor A detecting device which is part of a *transducer*. It will extract energy from the measured medium in the process of measurement.

separation The treatment of fuel and lubricating oils to remove water and sediment, usually by the action of gravity in a heated tank and then centrifuging.

separator A device which removes oil from oil–water mixtures. The operating action is related to the gravity differential between the two liquids for separation and may also include a coalescing filter where extreme cleanliness is required.

series connection An electrical connection of *cells* in a *battery* where the positive and negative terminals of adjacent cells are connected and their voltages are added.

series wound A d.c. electrical machine whose field windings are in series with the armature.

service boat A ship which has been designed and constructed, or adapted, for the support the offshore drilling operations. It may be used for towing, transporting supplies, anchor handling or other duties and is sometimes called a *supply boat*. See Figure S.8.

service tank A *tank* which is supplying clean, treated fuel oil directly to the engine.

servomechanism An automatic monitored *kinetic control system* which includes a power amplifier in the main forward path.

servomotor The final control element present in a *servomechanism*. It is the motor which receives the output from the amplifier element and drives the load.

180

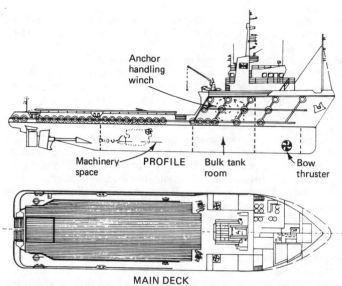

Figure S.8 A service boat

set value   The *command signal* which is supplied to a regulating system.

settling tank   The *tank* into which oil is transferred from the bunkers or double bottom tanks. The oil is then heated and water and impurities are allowed to settle before further fuel treatment takes place.

settling time   The time taken for the *index* of an instrument or the *controlled condition* of a system to reach and remain within a specified *deviation* from its final steady value, after an abrupt change.

sewage treatment   The use of either biological or chemical processing to render sewage suitable for discharge in permitted areas or to a shore reception facility. The discharge of untreated sewage is banned by legislation.

sextant   An instrument with mirrors and a pivoted arm which can move through a graduated arc of a sixth of a circle. It is used by navigators to determine a ship's position by measuring the altitude of the sun or stars.

shaft   A circular section rod or bar which transmits rotary motion, e.g. propeller shaft.

shaft alignment   The positioning of all the propulsion shafting in line from the engine. The alignment is checked during installation by a taut wire, a sighting telescope or a laser beam. Further static and dynamic alignment can be done when the vessel is in service.

shaft bearing   A *bearing* which supports the propeller shafting. The aftermost bearing or *plummer block* has top and bottom bearing shells. All other shaft bearings have only a lower half bearing shell.

181

shaft bossing A plated structure, stiffened by frames, which encloses the shafts of a twin or multiple screw vessel. It extends from where the shaft emerges and ends in the *shaft bracket*.

shaft bracket The casting at the end of the *shaft bossing*.

shaft coupling A connection between two lengths of propeller shafting.

shaft generator A *generator* which is driven by the main engine or the propulsion shafting. The drive may be by belt, gears or other arrangement since constant speed operation is required.

shaft power The power available at the output shaft of an engine. It is usually measured by a *torsion meter* on or around the propulsion shafting.

shaft tunnel A watertight structure used to surround the propeller shafting when the machinery space is not fully aft. It protects the shafting in way of the cargo holds and provides watertight integrity should the shaft seal fail. The forward end is fitted with a sliding *watertight door* and an escape route is provided from the after end.

shafting The lengths of propeller shaft between the engine, or gearbox, and the propeller.

shale shaker A sloping, vibrating, close mesh screen through which the *drilling fluid* is passed on its way to the mud tank. Any drill cuttings and large solids are thus removed.

shear deflection The movement of a beam resulting from *shear stress* during bending. Vibration frequency is related to deflection and if shear deflection is appreciable it will influence the frequency.

shear force A force tending to cause displacement of a transverse plane due to unequal loading. The distribution of load along a ship and the upthrust or buoyancy are unequal at various points and cause such shear forces to exist.

shear stress The *shear force* intensity per unit area of cross section, which varies across the section. It is a consideration in the structural strength of ships which are subjected to longitudinal bending.

sheath A covering fitted over the *insulation* of an electric cable to protect against heat, oil and chemicals. It must also be tough and flexible, e.g. hypalon.

shedder plate A sloping plate fitted within the trough of a corrugated bulkhead to enable a steady downward flow of bulk cargoes during discharge and easier hold cleaning.

sheer The curvature of the deck in a longitudinal direction. It is measured between the deck height at midships and the particular point on the deck. See Figure P.7.

sheer profile A set of sections of a ship's form obtained by the intersection of a series of vertical planes parallel to the centreline of the ship with the outside surface. They are part of the *lines plan*. See Figure L.3.

sheer strake The strake or plate of side plating nearest to the deck. It is usually increased in thickness or a higher tensile steel is used because of the high bending stresses experienced.

shell and tube heat exchanger A *heat exchanger* in which a tube bundle or

182

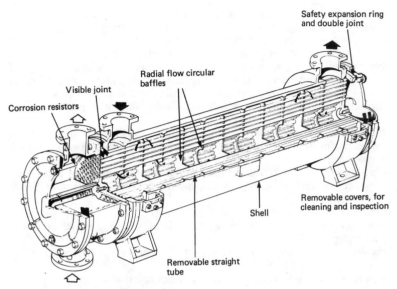

Safety expansion ring
and double joint

Radial flow circular
baffles

Visible joint

Corrosion resistors

Removable covers, for
cleaning and inspection

Shell

Removable straight
tube

Figure S.9 A shell and tube heat exchanger

stack is fitted into a shell. The end plates are sealed at either end and cover plates over them enable the entry and exit, usually of sea water. The liquid to be cooled enters through the shell, circulates over the tubes and exits through the shell. See Figure S.9.

shell bearing A semicircular thin steel shell which is lined with a suitable low friction bearing metal. A pair will be used and they are usually designed such that they cannot be incorrectly fitted.

shell displacement The displacement of the shell plating. The wetted surface area is first found and is then multiplied by the shell thickness and the density of sea water to obtain a value in tonnes which can, if required, be converted into a force.

shell expansion A plan which shows the position and thickness of all the plates which comprise the *shell plating* of a ship. The extent of each prefabricated unit of the ship will also normally be shown.

shell plating The strakes or plates which form the watertight outer skin of a ship.

shelter deck A term applied to a superstructure deck which extended the length of a ship and had at least one tonnage opening. It was a means of reducing a ship's tonnage measurement.

shifting boards Wooden planks which are used to partition holds to prevent the movement of loose bulk cargoes. They are fitted fore and aft on the centreline in way of the hatches.

shim A thin strip of metal which is used to adjust the clearance between mating parts, e.g. the two halves of a big end bearing.

ship (1) A sea-going or coastal vessel which is not propelled by oars. (2) The act of putting on board or sending goods as cargo by ship.

ship broker An agent acting for the shipowner who obtains cargo and passengers and deals with all matters relating to the ship in port. A broker may also deal with chartering or the buying and selling of ships.

ship canal A canal through which ocean-going ships may pass, e.g. Suez Canal.

ship's official number An official number which is allocated to a ship upon registration. It is cut, punched or stamped into a main beam or hatch coaming.

ship's papers The *Certificate of Registry, manifest,* crew list, *log book,* crew agreement, *bills of lading,* any existing *charter party* and a *Bill of Health* where required.

shipment A quantity of goods sent by sea.

shipowner The person or company which is the officially registered owner of a ship or ships.

shipping (1) A collective term relating to all aspects of marine transport. (2) The transport of goods by sea.

shipwreck A ship which has been lost or destroyed at sea.

shipyard The place where ships are built or repaired.

shock valve A *relief valve* fitted between the hydraulic pipelines of a steering gear system. It will open if pressure in the system rises by 10% due to a loading on the rudder. It will, for example, open and allow the rudder to move freely if it is struck by a heavy sea. The *hunting gear* will return the rudder to its previous position.

shoe The part of a brake which is brought into contact with a drum to produce a frictional force which will stop the movement.

shore supply An electricity supply obtained from the shore mains when a ship is in dock undergoing repairs.

short circuit A fault condition where a low resistance connection occurs between two points in a circuit. A large current flow will usually occur.

short ton A ton of 2000 pounds.

shot-blasting A cleaning process for metal plate in which dry metal shot or steel grit is projected at the surface.

show See *strike.*

shrinkage The temporary falling of boiler water level as incoming colder feedwater causes the collapse of steam bubbles. It is a phenomenon observed in a watertube boiler as the control system corrects for an increase in load.

shrouded propeller A *propeller* fitted within a duct or nozzle in order to improve its efficiency. It is often used in small ships such as tugs.

shrouding Strips of metal which are attached to the outer edge of a number of blades in a steam turbine.

shunt wound A d.c. electrical machine whose field windings are in parallel with the armature.

shuttle valve A *valve* which moves from end to end of the valve chest. It controls the supply of steam to the cylinder of a double-acting single cylinder *reciprocating pump.*

SI units A system of coherent metric units which has been increasingly used since 1960. See Appendix.

side loading The entry of cargo through a door in a ship's side.

side rod A rod which connects the transverse beam of an *opposed piston engine* to the side crosshead. See Figure O.2.

side scuttle See *porthole.*

siemens The unit of electrical *conductance.* It is a reciprocal *ohm.*

signal A physical quantity used to transmit information between one element of a control system and another.

signal letters The sequence of letters allocated to a ship which are used to identify it in all communications between ship and shore.

signal processing The manipulation of information contained in a signal by modulating, demodulating, mixing, gating, computing or filtering.

silica gel A granular form of hydrated silica. It is used as a *desiccant.*

silicon A non-metallic element which is the main constituent of many rocks and clays. It is a *semiconductor.*

silicon controlled rectifier A three junction *semiconductor* device. It is also known as a *thyristor.*

silicone A *silicon* compound of oxygen and other elements which produces an oil with a low melting point and a *viscosity* which changes little with temperature. It is used as a lubricant and a hydraulic liquid. Solid silicone compounds are often used as electrical insulators.

sill (1) The plating beneath an opening such as a door. (2) The upper edge at the bottom of the entrance into a dock.

Simpson's rule A mathematical formula which is used to represent a curved surface and enable the determination of an area beneath it. It has considerable applications when dealing with curves related to a ship's form.

simulation The representation of individual physical items in a system by computing elements. The interconnection of the computing elements is the same as in the simulated system and enables it to be tested.

singing A high pitched note given out by a *propeller* just before *cavitation* begins. It is due to *vibration* of the blades.

single bottom structure A construction method used in tankers and some small vessels where the bottom shell plating is stiffened by plate floors and longitudinal stiffeners.

single buoy mooring Another term for *single point buoy mooring.*

single phasing A fault condition in an *induction motor* when one phase in a three phase circuit becomes open circuited. Excessive currents will occur in all the windings.

single point buoy mooring A *buoy* located in deep water to enable a tanker

185

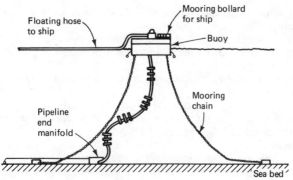

Figure S.10 A single point buoy mooring

to moor and load oil, whilst swinging on the mooring. Special types exist such as the *spar buoy*, the *single anchor leg mooring* and the *exposed location single buoy mooring*. See Figure S.10.

single well oil production system The use of a specially designed or a converted oil tanker with appropriate equipment to act as a *production platform* for a marginal oil field.

skeg A vertical projection fitted at the bottom of the after end of a ship. It may serve to support the lower end of the rudder or the ship itself when in drydock.

skew The offset of a *propeller* blade from the vertical in the plane of rotation. It is always a distance in the direction opposite to rotation. See Figure P.8.

skin friction The resistance component of a ship which is due to the roughness of the hull plating. See *frictional resistance*.

slamming The effect resulting from the rise and fall of the forward end of a ship when it is *heaving* and *pitching*. Large forces can thus act on the forward bottom plating. It is also called *pounding*.

slave controller A *controller* which is used in a *cascade control system*. It has a variable *desired value* that is set by a master controller.

slenderness ratio The ratio of the length of a column to the radius of gyration of its section. It is an important factor when considering the resistance to *buckling* of structural members of a ship.

slewing To rotate about a vertical axis in a horizontal plane.

slide valve A rectangular *valve* which slides over ports leading to a double-acting steam engine cylinder. Steam acts on the back of the valve and the exhaust port is open to the inside face of the valve. As it moves steam enters one side of the cylinder and is able to exhaust from the other.

sliding foot A supporting foot or palm which permits linear movement as expansion takes place on, for example, a boiler casing or a steam turbine.

slimness An indication of the form of a ship using the ratio of length divided by the cube root of the volume of displacement. Froude and others have

186

used various coefficients to represent this.

sling A loop of rope or wire which is used in the hoisting of a heavy item. It may also describe a length of wire with a loop at each end.

slip (1) The difference between the actual distance travelled by a ship and the theoretical distance given by the product of the propeller pitch and the number of revolutions. It is usually expressed as a percentage and can have a negative value if a current or following wind exists. (2) The inclined ways upon which a ship is built and then launched. (3) To separate from by unshackling a link of chain, e.g. slip an anchor. (4) The ratio of the difference between synchronous speed and actual speed divided by synchronous speed, for an *induction motor*.

slip-coupling An electromagnetic *coupling* in which the parts are not in physical contact. Two electromagnets are formed and due to a small speed difference, or slip, between the driving and driven members power transmission takes place. This coupling arrangement will damp out *torsional vibrations*.

slip ring A ring fitted on a rotating shaft to provide an electrical connection between a moving and a fixed conductor by the use of one or more *brushes*.

slipper pad pump See *swashplate pump*.

slipway The inclined ways upon which a ship is built and then launched.

slop tank A *tank* which is used to store slops or the oil–water mixture resulting from tank cleaning on an oil tanker. A pair are fitted at the after end of the cargo tank section.

slow-speed diesel A main propulsion engine which rotates at about 80–120 rev/min. It can thus be directly connected to the propeller shaft without the use of a gearbox. See Figure S.11.

slow steaming The operation of a ship at a lower than normal speed in order to save fuel on ballast voyages or when fuel is expensive.

sludge The *residue* from any oil separation process which is usually a mixture of water, solid material and high viscosity oil.

sluice valve A large *valve* in which a rectangular or circular gate slides across the opening. It has been used in oil tankers to permit gravity flow from tank to tank, with the valve being operated from the weather deck.

smoke detector A specific type of *fire detector* which is fitted in a space where smoke will indicate the presence of fire.

snifting valve A vent valve used in hydraulic systems to release any trapped gases.

soft-nosed stem A stiffened, radiused, plated structure extending from the load waterline to the forecastle. It provides a collapsible region at the bows to minimize impact damage in the event of a collision.

software A general term for computer *programs*.

soil pipe A discharge pipe from a water closet or urinal.

solar gear A system of *epicyclic gears* in which the sun wheel is fixed, the annulus and planet carrier rotate, and the planet wheels rotate about their own axis. See Figure E.3.

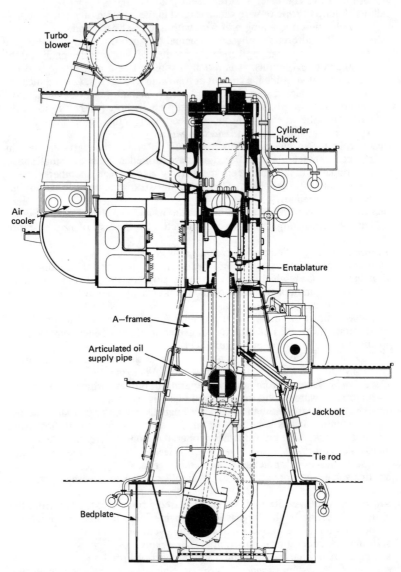

Figure S.11 A slow-speed diesel

Turbo blower

Cylinder block

Air cooler

Entablature

A—frames

Articulated oil supply pipe

Jackbolt

Tie rod

Bedplate

SOLAS An acronym for the Safety of Life at Sea Convention of the *International Maritime Organization.* The 1972 Convention is presently in force.

solenoid A coil of wire with a long length in relation to its diameter. The coil is tubular and designed to act as an *electromagnet* in order to move an iron bar along the axis of the coil.

solid injection The injection of a liquid into an enclosed space by the use of hydraulic pressure only, e.g. fuel injection into an engine cylinder.

solid state A device, circuit or system whose operation depends upon any combination of electrical, magnetic or optical phenomena within a solid, i.e. no movement takes place.

sonar SOund NAvigation Ranging. A device which is used to detect and position underwater objects, e.g. submarines.

sootblowers Equipment which is used to remove the products of combustion from boiler tubes. Rotating nozzles may be used in low temperature regions whilst retractable lances with holes along their length are used in high temperature regions. Steam is blown in most cases although compressed air may be used.

sound-powered telephone A telephone which requires no external power source. They are fitted to communicate between the bridge and the machinery space and between the bridge and the steering gear.

sounding pipe A pipe which leads down to almost the bottom of an oil or water storage tank to enable the depth of liquid to be measured by a sounding tape. A striking pad must be located on the tank bottom beneath the pipe.

source rock The rock from which oil or gas originally formed. It is usually a form of organic-rich sedimentary rock.

spaces The areas of a ship which are used for particular purposes, e.g. machinery spaces.

span The input signal range of an instrument that corresponds to the designed working range of the output signal.

spar buoy A large *single point buoy mooring* which is usually manned and also provides an oil storage facility.

special area A sea area in which particular requirements regarding oil pollution must be met to protect the environment. They are named in the *MARPOL 73/78* Convention, e.g. Mediterranean Sea and Red Sea.

special survey A rigorous examination of a ship's structure and fittings for *classification* purposes. It takes place every four years.

specific fuel consumption The amount of fuel used in unit time to produce unit power. The unit is kg/kW h.

spectacle frame A large casting which projects outboard from the ship and supports the ends of the propeller shafts in a *twin screw vessel.* The casting is plated into the surrounding shell. See Figure S.12.

spectral density A mathematical function of which any area under its graph, when multiplied by a suitable constant, represents the wave energy in that incremental band of frequencies. See *wave spectra.*

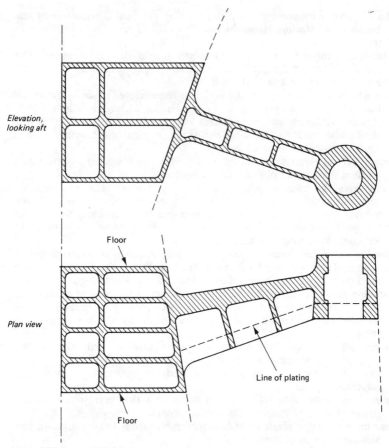

Figure S.12 A spectacle frame

speed (1) The ratio of the linear distance travelled by a body to the time taken. The usual unit is metres per second but use is also made of the knot which is one nautical mile per hour. (2) Where an angular velocity is considered this is usually given as revolutions per second or per minute.

speed/length ratio The quantity $v/\sqrt{l}$, where v is speed and l is length. It is used to define whether a ship is fast or slow when considering resistance.

speed of advance The speed of a ship relative to the wake.

speed through the water A ship speed measurement which is obtained by the use of some form of *log*. It may be different to a speed related to land because of currents and tides.

speed trials Test runs which are made by a newly completed ship to determine the speed through the water, the engine revolutions per minute and the power developed. A measured course is used and water of reasonable depth along a coastline.

spike A transient of short duration which is an unwanted portion of a pulse with a large amplitude, e.g. voltage spike. It can damage sensitive electronic equipment.

spiked cargo A crude oil cargo which contains a quantity of liquefied natural gas.

spindle A part of a machine which rotates, e.g. lathe spindle, or a pin upon which another part turns.

spiral gears A toothed gear which connects two shafts whose axes are at an angle and do not intersect. The teeth are formed in a spiral and engage in the same way as a *worm gear.*

splash proof An *enclosure* which protects electrical equipment from drops of liquid or solid particles which fall onto it or travel in a straight line at any angle not greater than 100 degrees from the vertical.

spline shaft A *shaft* with longitudinal grooves around its periphery to produce a series of narrow keys or splines. An internally splined member can thus fit over the shaft and be mechanically coupled. Axial movement may be permitted in some arrangements.

split bearing A *bearing* which is produced, usually in two halves, to enable removal by removing a cap or part of the enclosure and sliding the bearing out.

split range control A form of multi-loop control where the output from a single *controller* is split into two or more ranges to operate correspondingly more *correcting units.*

split windlass The use of a separate *windlass* for each anchor.

spontaneous combustion A fire caused by a chemical reaction within a substance. Coal and some bulk solids may oxidize and suddenly begin to smoulder or burn due to spontaneous combustion.

spool valve A *valve* arrangement where a number of spools or bobbins on a spindle slide in a cylinder to cover and uncover various ports. It is often used in hydraulic control systems.

spread The arrangement of *seismometers* when undertaking a seismic survey.

spreader A lifting arrangement which is used to separate the *slings* or wires in order to avoid crushing the load.

spreading function A mathematical value introduced into *wave spectra* calculations in order to take into account the different directions of wave travel.

sprinkler system An automatic water spraying system which protects against fire. A quartzoid bulb in a sprinkler head will shatter as a result of temperature rise and then pressurized water will be sprayed into the space. One or more sprinkler heads will be fitted in the various protected spaces.

191

sprocket An alternating arrangement of teeth and spaces around the rim of a wheel or drum which will engage the links of a chain.

spud in To begin drilling a new *borehole.*

spur gear Gear wheel arrangements where the teeth are parallel to the axis of rotation.

spurling pipe A heavy plate pipe which is fitted at the entrance to the *chain locker* to lead the anchor cable in and out. A solid round bar chaffing ring is fitted on the lower edge inside the chain locker. See Figure C.5.

square root extractor A device which is used in measurement or control systems to provide an output signal which is the square root of the input.

squat The mean increase in draught, i.e. sinkage, plus any contribution due to trim as a result of a vessel travelling at a considerable speed. It is an important factor when the depth of water is less than 1.5 times the draught.

squirrel cage motor An *induction motor* which uses three separately phased windings in the stator to produce a rotating magnetic field. The rotor is a cage formed from conductors in which an e.m.f. is created by electromagnetic induction. The current-carrying conductors in the cage result in a motor effect.

stability (1) For an *instrument* this means that repeated readings, taken over long periods, under defined conditions, give the same results. (2) A *control system* is considered stable if the response to an impulse input approaches zero as time approaches infinity. (3) A *ship* is stable if, when floating in still water, it is displaced by some external force and moves away from its equilibrium position but returns to that position when the force is removed. See *metacentric height.*

stabilizer Any device used to reduce the *rolling* motion of a ship. It may be passive, e.g. bilge keel, fixed fin or a tank system, or active, e.g. moving fin or a controlled tank system.

stabilograph An instrument, used in an *inclining experiment,* which records the angle of *heel* to a base of time.

stack controller A pneumatic *controller* which is a constructed of several diaphragm chambers mounted one above the other.

stagnation point A position of zero velocity in the flow pattern around a cylinder or aerofoil.

stalling (1) The breakdown of a lift producing flow over an aerofoil section. (2) The stopping of an engine due to the sudden application of a load or braking force.

stanchion A vertical metal post which forms part of a handrail or a protective barrier.

standard charter party forms Standard forms which may be used when drawing up a *charter party.* Various suggested terms or clauses may be modified as required by the contracting parties.

standard fire test The exposure of a material specimen, in a test furnace, to a particular temperature for a certain period of time.

Standards of Training, Certification and Watchkeeping for Seafarers,

1978 An *International Maritime Organization* Convention which has established internationally acceptable minimum standards for crews.

standby vessel A ship which remains on location near an offshore installation and can in an emergency, accommodate all persons from the installation and provide first aid treatment, if required. UK regulations require such a vessel in all safety zones.

star connection The joining of three conductors or windings such that they meet at a common point.

star-delta starting A means of reducing the starting current of an *induction motor* by initially connecting the stator windings in star. Once the motor is running the windings are connected in delta.

star gear A system of *epicyclic gears* in which the planet carrier is fixed, the annulus and sun wheels rotate and the planet wheels rotate about their own axis. See Figure E.3.

starboard The right-hand side of a ship when facing forward.

starter An electric controller which is used to start a motor, accelerate it up to normal running speed and, when necessary, stop it.

starting air compressor The *air compressor* which supplies compressed air at high pressure for starting the main propulsion diesel engine.

starting air receiver The storage cylinder or 'bottle' for high pressure compressed air which is used for starting the main propulsion diesel engine.

starting air system The arrangement of equipment, piping, controls, etc., from the starting air compressor to the air starting valve in the engine cylinder.

starting current The *current* drawn by a motor when it is started and running up to normal speed.

static (1) At rest or in a state of equilibrium. (2) An electrical charge which may be induced by friction or atmospheric effects.

static delivery head The vertical distance from the pump to the delivery liquid level or the highest point in the system.

static excitation The use of *transformers* and *rectifiers* to produce the series and shunt components of an a.c. generator's field windings.

static frequency converter A device used with *shaft generator* systems to provide a constant output voltage and frequency. The shaft generator voltage output is rectified into a variable d.c. voltage and then inverted into a three-phase a.c. voltage. A feedback system within the oscillator inverter ensures a constant output voltage and frequency.

static suction lift The vertical distance that a liquid must rise before it enters the *pump*.

statical stability The ability of a vessel to return to the upright position when inclined by some force. See *curve of statical stability*.

stations The positions resulting from ten equally spaced divisions along a ship's length between the forward and after perpendiculars. The after perpendicular is numbered zero and the forward perpendicular is numbered ten.

193

stator The stationary part of an electrical machine including any magnetic parts and their windings.

stays Wires or ropes from the deck to the head of a *mast, samson post* or *boom* to provide support or prevent movement.

steady state The final condition that a physical quantity of a system reaches when the effects of all external disturbances have ceased.

stealer strake A single wide plate which replaces two narrow plates in adjacent strakes of a ship's plating.

steam Water which has been converted into a vapour by the application of sufficient heat energy to make the liquid change state. It may be *wet steam*, dry saturated steam, or *superheated steam*.

steam drum The upper drum of a two or multidrum boiler in which steam is released from the boiling water. It is maintained half full of water as the normal operating level.

steam pump Any *pump* driven by steam.

steam ship A ship whose propulsion machinery is powered by steam. It is given the designatory letters SS before the name.

steam-to-steam generator A form of *boiler* where the heating medium is high pressure steam which passes through coils in the boiler drum. Low pressure steam is produced for auxiliary purposes. The high pressure steam system thus avoids any possible feed contamination from auxiliary systems. See Figure S.13.

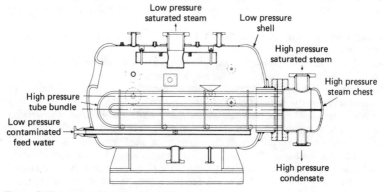

Figure S.13 A steam-to-steam generator

steam trap A fitting in a pipeline which will only permit the passage of condensed steam, i.e. water.

steam turbine A machine which converts the energy in steam into mechanical power. High pressure steam from a boiler is expanded in nozzles to create a high velocity jet. The jet of steam is directed into blades mounted on the periphery of a wheel, which brings about rotation of the rotor.

194

steam whistle A whistle which is operated by steam and used to give audible signals. It must have a range of several miles.

steel An alloy of carbon and iron with other metals added to improve the properties, reduce the heat treatment necessary and provide uniformity in large masses of the material. See also *higher tensile steel* and *mild steel*.

steel section A length of mild steel bar of standard shape, e.g. offset bulb plate, equal angle, channel. They are used as stiffeners for steel plate. See Figure S.14.

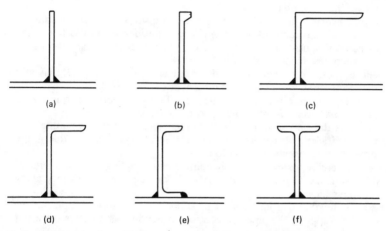

Figure S.14 Steel sections: (a) flat plate, (b) offset bulb plate, (c) equal angle, (d) unequal angle, (e) channel, (f) tee

steering gear The complete system which provides a movement of the rudder in response to a signal from the bridge. It consists of the control equipment, a power unit and a transmission to the rudder stock.

stellite An alloy of cobalt, chromium, tungsten and molybdenum which is very hard and used for cutting tools and the coating of valves.

stem The most forward part of the hull structure. It consists of a stem bar from the keel to the load water line and a stiffened plated structure up to the forecastle. See *soft-nosed stem*.

stem head The upper part of the *stem*.

step-down transformer A *transformer* in which the secondary voltage is lower than the primary voltage.

step function response The transient response resulting from an input signal or disturbance which is a sudden occurrence or step function.

step-out well A *well* drilled some distance from a discovery well. It may be for appraisal or development depending upon the drilling programme requirements.

stepper motor A d.c. motor which is used with computer controlled servomechanisms. An input pulse will cause rotation of the shaft through a fixed angle or step.

stern The after end of a ship.

stern frame A large structural item at the after end which houses the *stern tube* and, on single screw ships, the rudder supporting arrangement. It is welded to the after end shell plating and the keel. See Figure B.1.

stern post See *rudder post*.

stern tube A hole in the *stern frame* through which the *tailshaft* protrudes. It houses the *stern tube bearing*.

stern tube bearing The final bearing in the transmission system leading to the propeller. It supports the *tailshaft* and a considerable proportion of the propeller weight. In addition it acts as a gland to prevent the entry of sea water into the machinery space.

stevedore A dock worker who loads or unloads a ship's cargo. It may also refer to the person in charge of a number of dock workers.

stiff A ship which rolls in a jerky manner with a short period of roll, due to a large *metacentric height*. The opposite of *tender*.

stiffener A flat bar, section or built-up section used to stiffen plating.

stiffness coefficient When considering a *kinetic control system* this is the force or torque per unit deviation.

still water bending moment The bending moment acting on a ship lying in still water due to the uneven distribution of weight and buoyancy along the ship's length.

stinger A flexible or rigid support positioned at the stern of a *pipe-laying vessel*. It supports the pipe and controls its bending as it enters the water and sinks down onto the sea bed.

stocks The timber structure which supports a ship under construction.

stoichiometric mixture The chemically correct mixture of substances in order to bring about a reaction or create an alloy.

stop valve A *valve* which, when closed, will shut off a supply of fluid, e.g. main steam stop valve fitted on the boiler drum.

store (1) To save information in a computer memory. (2) A compartment in which items are stored, e.g. paint store.

stowage plan An outline plan of a ship's holds showing the cargo disposition, descriptive marks and destination. It enables preplanning at the port for the cargo to be discharged.

strain The deformation of a material due to *stress*. It may be compressive if the force tends to shorten the material or tensile if it tends to lengthen it.

strain gauge A length of fine wire whose resistance will change if it is displaced. The gauge is bonded to the item whose displacement is to be measured and is connected into a measuring bridge. See *Wheatstone bridge*.

strainer A coarse *filter* to remove larger contaminating particles from a system.

strake A continuous line of plating extending fore and aft over the length of a ship.

stranded A ship which has come into contact with a rock, sandbank or the shore and remained stationary for a period of time.

streamlined A structure whose shape has been smoothed such that when a fluid is flowing over it the resistance is minimal, e.g. appendages on a ship's hull.

strength calculation The determination of the structural strength of a ship or floating structure, considering both static and dynamic loading.

strength deck The *deck* which is considered as the uppermost part of the hull in strength calculations.

stress The force acting on a unit area of a material.

stress corrosion The combined effects of *tensile stress* and a corrosive environment which can result in intercrystalline cracking of a metal and early, unexpected, failure. It is also known as *caustic cracking* when referring to boilers, particularly those of riveted construction.

strike A successful drilling operation which produces evidence of hydrocarbons. It may also be called a *show*.

stringer A horizontal stiffener fitted along the ship's side or a longitudinal bulkhead, in order to provide strength and rigidity.

stringer plate The outboard strake of plating on any deck.

strip down To dismantle into component parts in order to examine or repair an item of equipment or machinery.

stripping pump A *pump* which draws the remnants of an oil cargo from the hold and discharges it to the deck manifold.

stroke The distance travelled by a piston between *top dead centre* and *bottom dead centre*, i.e. the maximum of travel.

structural failure This is considered to occur when all the material of a bending structure, e.g. a ship, has reached its *yield stress*.

strum A wide mouthed fitting on the end of a suction pipe close to the bottom of the tank or well to improve suction and minimize the entry of air into the pump.

strum box See *mud box*.

strut A light, structural member or long column, which is designed to accept an axial compressive load.

stud welding The use of a gun-type device for affixing small metal pins by electric *arc welding*.

stuffing The filling of *containers* with cargo.

stuffing box A space, in a casting or structure, around a shaft which is filled with packing to make a liquid or gas-tight seal. The packing is compressed by an adjustable gland ring. See Figure S.3.

Stulken derrick A patented type of *heavy lift derrick* which is positioned between two outwardly raked, tapering, tubular columns. See Figure H.3.

sub-sea completion The mounting of oil flow controlling equipment, i.e. a *christmas tree*, on or beneath the sea bed to complete the *well*.

sub-sea production system The mounting of most of the production equipment on, or under, the sea bed. Servicing may be by diver or remote operated vehicle.

subassembly   Several pieces of steel plate which have been welded together to form a two-dimensional part of a ship's structure. It may weigh up to five tonnes and examples would be *transverses,* minor bulkheads and *web frames.*

submersible   (1) An underwater vehicle suitable for deep diving in order to undertake survey, installation, repair or recovery work related to oil rigs or submarine cables. (2) A piece of equipment which can operate when underwater, e.g. submersible bilge pump.

suction   The drawing in of a gas or liquid to a pipe or pump as a result of a reduction of pressure therein.

suction pressure decay   The reduction in *net positive suction head* of a pump as the flow rate increases.

sulphur content   A fuel oil property which is important since sulphur is considered a cause of engine wear. A maximum limit, expressed as a percentage by weight, is usually included in specifications.

summing amplifier   A device, in a *control system,* which will add algebraically the signals input to it and amplify the summed output signal.

sump   The *crankcase* of an engine, when it is used as a tank to store lubricating oil.

supercargo   A person engaged on board a vessel to oversee the stowage and care of the cargo. This person is usually employed by the vessel's charterers and attends to their interests with respect to the cargo without interfering with the management of the vessel.

supercharging   Introducing a larger mass of air, i.e. the charge, into an engine cylinder by blowing it in under pressure. See *turbocharging.*

superconducting   A phenomenon occurring in certain metals, alloys and compounds where, when cooled to almost absolute zero, their resistance becomes almost negligible. Motors using this phenomenon have been developed.

superheated steam   *Steam* which has been removed from the boiler drum and further heated in a *superheater* to a temperature higher than the saturated temperature corresponding to the boiler pressure.

superheater   A bank of tubes through which saturated steam from the boiler drum is passed. Hot gases from the boiler furnace pass over the tubes, dry the steam and then superheat it.

superintendent   A senior staff member of a shipowning or ship managing company who will have a particular responsibility for one or more vessels or aspects of their operation, e.g. engineering or marine superintendent.

superstructure   The part of a ship's structure which is built above the uppermost complete deck and is the full width of the ship.

superstructure efficiency   The ability of a superstructure to diffuse or accept forces associated with the longitudinal bending of a ship. It is largely dependent upon the ratio of the superstructure length to its transverse dimension.

supply vessel Another term for *service boat.*

surface condenser A steam *condenser* of shell and tube construction. Sea water circulates through the tubes to condense the steam.

surge A sudden large flow of short duration, e.g. current, voltage, liquid or gas.

surging The fore and aft linear motion of a ship in the sea. See Figure R.6.

survey A detailed examination of a ship's structure and fittings in order to verify a particular standard of fitness or condition. It may take various forms for a ship, e.g. annual survey, special survey, etc.

Surveyor See *Lloyd's Surveyor, Department of Transport Surveyor.*

susceptance The component part of admittance which is due to *inductance, capacitance* or both. It is the reciprocal of *reactance.*

swashplate (1) A longitudinal or transverse, non-watertight, tank division whose function is to minimize the sloshing action of the liquid contents as the ship rolls or pitches. (2) A circular flat plate which is obliquely mounted on a rotating or fixed shaft.

swashplate pump An axial cylinder *variable delivery pump* used in hydraulic systems. The driving shaft rotates the cylinder barrel, swash plate and pistons. An external trunnion enables the swash plate to be moved about its axis which varies the stroke of the pistons in the barrel. See Figure S.15.

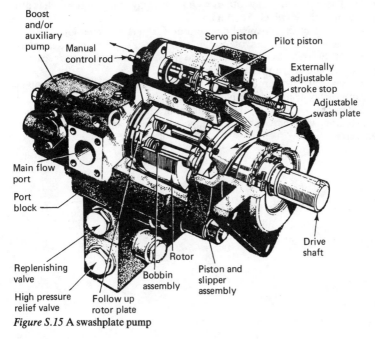

Figure S.15 A swashplate pump

199

SWATH An acronym for Small Waterplane Area, Twin Hull vessel. Twin torpedo-shaped hulls are fully submerged with streamlined fins or struts supporting the upper platform or deck. It is used for passenger carrying and research vessels because it provides a stable platform.

swaying The side to side linear motion of a ship in the sea. See Figure R.6.

swell (1) A rising and falling motion of the sea in fine weather. (2) The temporary rise in boiler water level after an increase in load. It is due to an increasing number of steam bubbles forming in the water as the drum pressure falls.

swept volume The volume of gas displaced by a *piston* when moving through its *stroke.*

swim The protruding part of a bow or stern structure which is below the load waterline.

switchboard A large single panel, or assembly of panels, on which are mounted *circuit breakers,* switches, various protective devices and instruments. The *bus bars* receiving the generator output will be at the back and distribution will take place via *circuit breakers.*

switchgear All switches and power interrupting devices and their associated equipment.

swivel A support piece for the *drill string* with provision for the injection of drilling fluid into the drill pipe. It attaches to the hook below the *travelling block.* The top of the *kelly* rotates within a bearing in this unit which is also called the rotary swivel. See Figure D.2.

synchro An electromechanical device used for data transmission.

synchronizing The process of bringing the *voltage, frequency* and *phase angle* of two electrical supplies into line in order that they can be paralleled and share the load.

synchroscope An instrument which indicates when two electrical supplies are in synchronism and can be paralleled.

system diagram A drawing of an electrical or other system in order to show the relationship between the component parts and how the system operates.

T

table See *rotary table*.

table of offsets A book containing the *offsets* of a ship. These provide manufacturing information for the various trades involved in ship production.

tachogenerator An a.c. or d.c. *generator* which provides an output voltage proportional to its rotational speed. It may be used to measure speed or as part of an automatic control system to regulate speed.

tachometer An instrument which measures the angular velocity of rotating machinery giving a reading in revolutions per minute.

tail pipe A straight pipe leading from a mud box to the bilge with no fittings on the open end.

tailshaft The final length of shafting to which the *propeller* is attached. It has a flanged face where it joins the intermediate shafting. The other end is tapered to suit a similar taper on the propeller boss. It is also called the *propeller shaft*.

tangent tube A form of furnace wall construction in a *boiler*. The various *waterwall tubes* are fitted close together and backed by *refractory*, *insulation* and the boiler casing.

tangentially fired boiler A *boiler* which has *burners* fitted at each corner of the furnace and so aligned that they fire tangential to a circle. This arrangement is considered to improve turbulence and the mixing of air and fuel.

tank A reservoir, constructed of steel plate, which is used for storing liquids. It may be built-in to the double bottom spaces. It will have a filling pipe, suction, sounding and vent pipes together with an access manhole for cleaning or inspection. See *deep tank, feed tank, saddle tank, service tank, settling tank*.

tank cleaning The use of automatic mechanical equipment to spray hot or cold water around the cargo tanks of an oil tanker in order to clean them. Fixed or portable machines may be used which enter the tank through circular openings in the deck. It has, to a large extent, been superseded by *crude oil washing*.

tank testing (1) The testing of a scale model of a ship's hull in order to obtain information about resistance and hence powering required by the full size ship. The model is towed in a *towing tank*. (2) The pressure testing of a tank by filling with water.

tank top The steel plating which forms the top of the double bottom.

tank type boiler See *auxiliary boiler*.

tanker See *oil tanker*.

tappet A sliding, cylindrical item which moves within a guide due to the action of a *cam* and in turn operates a *push rod* or valve stem in a four-stroke engine, i.e. a cam follower.

201

tapping points   The locations from which the pressure readings are obtained when using an *orifice plate* or similar device to measure fluid flow.

Tchebycheff's rule   A means of determining the area under a curve by using a particular spacing of the ordinates which depends upon the number of ordinates used. The area is then the sum of the ordinates multiplied by the length of the curve and divided by the number of ordinates.

technician   See *Engineering Technician.*

Technician Engineer   See *Incorporated Engineer.*

telegraph   A communication system between the bridge and the machinery space which indicates speed and direction. The handle on the bridge is moved to indicate the particular speed and direction and also rings a bell. When the engine room telegraph is moved to the same position the bell stops. The telegraph may also be an integral part of the machinery controls. See Figure T.1.

telemotor control   A hydraulic control system for *steering gear* using a transmitter, a receiver, pipes and a charging unit. The transmitter, which is built into the steering wheel console, is located on the bridge and the receiver is mounted on the steering gear.

teletype   A teletypewriter or teleprinter. It is a telegraph system where the message is fed in by typewriter and received by another typewriter which prints it out.

telex   A communications device operating through the telephone network using a terminal, which is a *teletype* machine, at each end. A written message is provided by the teletype machine.

tell tale   An indicator of position or alignment of a movable object.

temperature   A reference to hotness or coldness of a body. The *Celsius scale* of measurement uses 0°C as the temperature of melting ice and 100°C as the temperature of boiling water, both at atmospheric pressure. See *kelvin scale.*

tempering   A process which follows the quenching of steel and involves reheating to some temperature up to about 680°C. The higher the tempering temperature the lower the tensile properties of the material.

template   A full-size pattern from which an item can be made, e.g. a pipe or a steel plate.

temporary hardness   The bicarbonates of calcium and magnesium, in boiler feed water, which are decomposed by heat and come out of solution as scale forming carbonates.

tender   A ship with a small amount of *stability* which rolls easily and is slow to return upright. The opposite of *stiff.*

tenon   A projection at the end of a component which fits into a socket of similar shape to enable assembly and a means of securing.

tensile strength   The ability of a material to withstand *tensile stress,* i.e. a force attempting to lengthen the material, divided by the area of the material. See *ultimate tensile stress.*

tensile stress   See *tensile strength.*

tensile test   A test to determine a material's strength and ductility. A

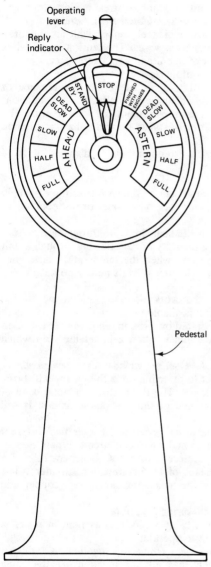

Figure T.1 A telegraph

203

specially shaped specimen of standard size is gripped in a testing machine and the ends drawn apart. The deformation of the specimen for different loads, i.e. the strain, can be found. A graph of stress against strain can then be drawn. See *modulus of elasticity, ultimate tensile stress.*

tension leg platform A floating concrete or steel *platform* which is anchored to caissons on the sea bed by numerous wires. The tension in the wires creates a stable platform which can operate in very deep water.

terminal A computer input and/or output device.

terotechnology A scientific study of total or life cycle maintenance of physical assets with a view to minimizing costs. It includes installation, commissioning, maintenance and replacement of machinery. The interrelationships between the different stages are also utilized to provide feedback for improvements.

territorial waters The waters which extend, usually a distance of three miles, from a coastline out to sea.

tesla See *gauss.*

test cock A *cock* from which a sample can be drawn for testing or simply visual examination, e.g. salinometer cock on a boiler or test cock on an oily water separator.

thermal conductivity A measure of the heat flow rate through a material. The specific value is given as the quantity of heat flowing through a unit area of unit thickness in one second when the temperature difference between the faces is one degree. The unit is watts per metre kelvin.

thermal cracking See *cracking.*

thermal efficiency The ratio of the work done by an engine to the mechanical equivalent of the heat available in the fuel.

thermal trip A device which is operated by a rise in temperature. It is used on circuit breakers, relays, etc., and is often a bimetallic strip which deflects when heated.

thermistor A type of *resistance thermometer* which uses a *semiconductor* material. An increase in temperature acting on a thermistor will bring about a large decrease in resistance. The thermistor is a small bead of copper to which is added cobalt, nickel and manganese oxides. It will measure from -250 to $+650°C$.

thermocouple A temperature measuring device which uses two different metals or alloys joined together to make a closed circuit. When the two junctions are at different temperatures, an e.m.f. is generated and a current flows, providing the means of temperature measurement. The choice of metals will determine the measuring range, e.g. copper and constantan, -200 to $+350°C$.

thermometer An *instrument* for measuring *temperature.*

thermometer pocket A narrow cavity in a pipe or machine in which a *thermometer* is placed to measure temperature.

thermopile A number of *thermocouples* connected together in series or parallel. The series arrangement has all hot junctions at the same temperature and all cold junctions at the same temperature. A very sensitive measurement is therefore possible. The parallel arrangement has the hot junctions at different temperatures and the cold junctions all at the same temperature. An average reading is thus obtained.

thermoplastic Any *plastic* material which can be softened under the action of heat or heat and pressure and then hardened by cooling, without any change in its properties, e.g. polyvinyl chloride (PVC), nylon.

thermosetting Any *plastic* material which can be moulded in a heated state, undergoes a chemical change on further heating and then sets hard, e.g. Bakelite, epoxy resin.

thermostat A device which responds according to *temperature*. It may act as a valve in, for example, the water cooling circuit of small i.c. engines. It may be used in electric circuits to operate as a switch.

thermostatic expansion valve An *automatically* operating expansion valve which responds to the temperature of the *refrigerant* at the outlet from the *evaporator*. It will open or close to vary the flow of refrigerant from the high to the low pressure side of the system. See Figure T.2.

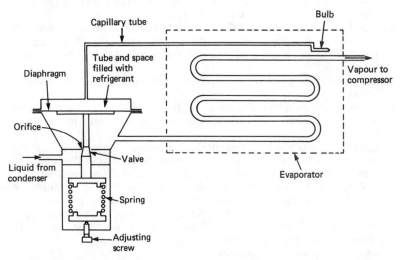

Figure T.2 A thermostatic expansion valve

thin shell bearing A *bearing* made up of a layer of low friction bearing metal bonded to a thin shell of stronger metal. This bearing is fitted into a bearing housing.

thixotropic A property of liquids and plastic solids which gives a high *viscosity* at low stress and a low viscosity at high stress. It is a useful property in paints as it enables the relatively easy application of a thick film.

205

three island ship A ship with a raised forecastle, a raised bridge deck amidships and a raised poop.

three-phase supply A combination of three circuits which are supplied by electro motive forces which are different in phase by one-third of a cycle, i.e. 120 degrees.

three-term controller A *controller* which provides *proportional, integral* and *derivative actions.*

threshold The value of input to a measuring instrument, or any system, below which no output change can be detected.

throttle valve (1) A *valve* which is used to control or regulate the flow of fluid through a pipe. (2) A *valve* which controls the supply of steam or fuel to an engine thus regulating either the speed or the output power.

Through Bill of Lading A *Bill of Lading* which covers the goods prior to loading or after their discharge.

thrust The component of force resulting from the action of a *propeller* which acts in line with the rotating axis.

thrust bearing An assembly of pivoting thrust pads which extend around the thrust collar of the *thrust shaft* on both sides and transfer the propeller thrust to the *thrust block*. See Figure T.3.

thrust block The complete assembly of *thrust shaft* and bearings which is solidly constructed and mounted onto a rigid seating. It will transfer the propeller thrust into the structure of the ship in order to bring about propulsion.

thrust power The product of propeller thrust and *speed of advance* of a vessel.

thrust shaft A short length of shaft with flanges at either end and a thrust collar in the centre. It may be manufactured as an integral part of some engines.

thruster A device which assists in docking, manoeuvring or positioning of a vessel which is moving at a low speed. Some form of propeller is used to move water either freely or in a duct. It may be fixed or controllable pitch and the complete unit may be retractable or exposed, fixed in position or able to rotate (azimuth). Some thrusters are used to provide main propulsion for small vessels, particularly when used in shallow draught regions. See *azimuth thruster, bow thruster, gill jet thruster*. See Figure T.4.

thyristor See *silicon controlled rectifier.*

ticket A *Certificate of Competency.*

tie rod A long bolt which is used to hold together the *bedplate, A-frames* and *cylinder block* of a slow-speed, two-stroke diesel engine. See Figure S.11.

tiller A casting or forging which is keyed to the *rudder stock* and used to turn the *rudder*. See Figure R.8.

timber load line A *load line* which is used only when a deck cargo of timber is carried. It is marked aft of the other load lines and preceded by the letter L.

time charter party A contract for the hire of a vessel for a specified time period. See *charter party.*

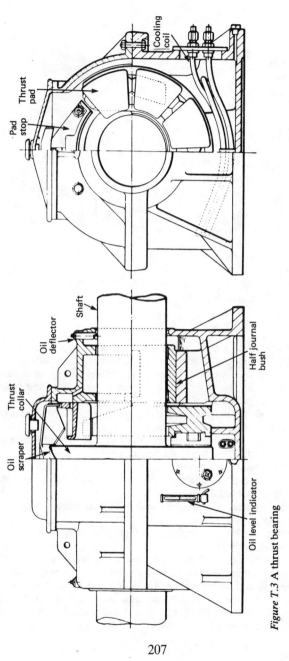

Figure T.3 A thrust bearing

207

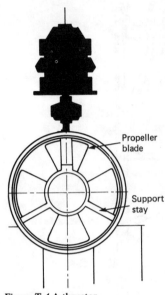
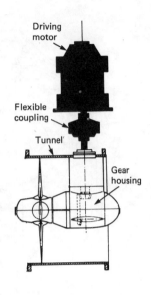

Figure T.4 A thruster

timing The moment in a cycle when a particular event takes place. It is described by the angle of the crank in relation either to top or bottom dead centre, e.g. inlet valve opens ten degrees before top dead centre.

timing diagram A diagram showing the opening periods of the inlet and exhaust valves of a four-stroke engine. See Figure T.5.

timing valve A *valve* which ensures the correct timing of a particular event, e.g. the injection of fuel into a cylinder.

tin A ductile, malleable metal that is resistant to corrosion by air or water. It is used as a coating for steel and also in various alloys.

titanium A light, strong, corrosion resistant metal which is used as the plate material in plate type heat exchangers. It is also used as an alloying element in various special steels.

tolerance The variation permitted in a dimensional size, position, form, etc., of a component.

tongs A wrench used for gripping, tightening and loosening of round objects, e.g. drilling pipe.

tonnage A measure of the internal capacity of a ship where 100 cubic feet or 2.83 cubic metres represents one ton. Two values are normally given for a ship, *net tonnage* and *gross tonnage*. The regulations which have governed the tonnage measurement of British ships are the Merchant Shipping (Tonnage) Regulations 1967(11). The International Convention on Tonnage Measurement of Ships, 1969, is now in force.

208

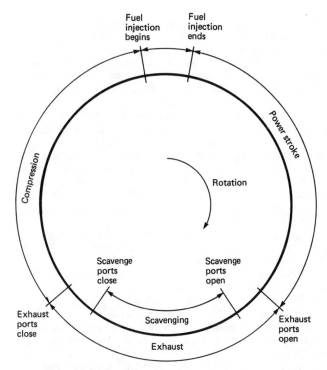

Figure T.5 A timing diagram

tonnage dues Charges levied on the basis of registered tonnage by a port authority.

Tonnage Mark A special *tonnage* draught mark on the side of a ship. If the mark is not submerged a lower value of tonnage may be claimed with respect to tonnage dues. Under the International Convention on Tonnage Measurement of Ships, 1969, this mark is not permitted.

tonne An *SI unit* of mass equal to 1000 kg.

tool (1) Any item used in a well for drilling related operations. (2) Any implement which enables the performance of a mechanical operation, e.g. manufacture, maintenance or repair. It may be operated by hand or some form of power.

tool joint Any junction or coupling in a *drill string*.

top dead centre The piston position when it is nearest to the cylinder head. The crank of the crankshaft will then be vertical.

torque The turning moment which is the product of a tangential force and the distance it acts from the axis of rotation. The unit is newton metre.

torque motor A motor of any type, but usually a.c., which does not rotate continuously, but is arranged to exert a torque opposing, for example, that of a spring or gyroscope.

torsion The *strain* set up in a member which is twisted without bending.

torsion box A rectangular cross-section girder running fore and aft on either side of a container ship. It extends from the ship's side to the cargo hold opening.

torsion meter An instrument which measures *torsion* in order to determine the torque acting on a rotating shaft. The *shaft power* of the engine driving the shaft can be found if the rotational speed is known.

torsional vibration The twisting of a shaft or structure about an axis in a cyclical manner due to a varying applied torque. If the frequency of the applied torque is the same as the natural frequency of the vibrated body, then resonance will occur. In an engine this would be a critical speed.

total base number An indication of the quantity of alkali, i.e. base, which is available in a lubricating oil to neutralize acids.

total head The difference in pressure between the suction and discharge branches of a pumping system, which is required to produce a flow of liquid. It is expressed as the height of a column of liquid.

total loss A ship which has been damaged to such an extent that it is of no value. See *constructive total loss*.

total static head The vertical height of a stationary column of liquid produced by a pump, measured from the suction level.

toughness A material property which is a combination of strength and the ability to absorb energy or deform plastically. It is a condition between brittleness and softness.

towing tank A tank of water in which a ship model is towed by a carriage travelling at various speeds. A dynamometer is used to measure the resistance of the model. Estimates of power requirements for the full-size ship can then be made on the basis of data obtained.

tramp ship A *general cargo ship* which is not on a scheduled run and generally seeks and carries its cargo anywhere.

transducer A device used for converting a signal or physical quantity of one kind into a corresponding physical quantity of another kind.

transfer function The equation, expressed as a Laplace transform, which represents the relationship between output and input for an element or a system.

transformer A device which generally consists of two windings on an iron core. An e.m.f. applied to the primary winding will induce an e.m.f. in the secondary winding in the ratio of the number of windings on the primary and secondary.

transhipment The transfer of cargo between ships or from a ship to its final destination.

transient Any change in conditions which occurs for a short period of time in, for example, an electric circuit or a control system.

transient response The time variation of the output signal that results when

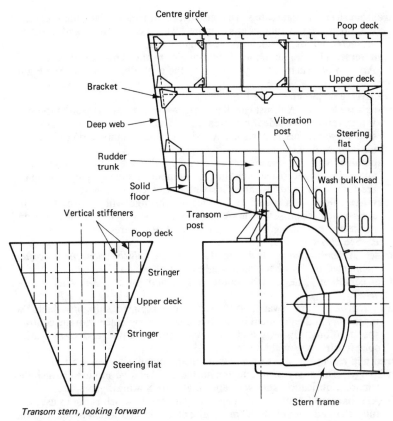

Centre girder

Poop deck

Upper deck

Bracket

Deep web

Vibration post

Steering flat

Rudder trunk

Solid floor

Wash bulkhead

Vertical stiffeners

Transom post

Poop deck

Stringer

Upper deck

Stringer

Steering flat

Stern frame

Transom stern, looking forward

Figure T.6 A transom stern

an input signal or disturbance of some specific nature is applied.

transistor An active *semiconductor* device which has three or more terminals.

transmission efficiency The ratio of *delivered power* to *shaft power*.

transmission system The various items which transfer the engine power to the propeller and the propeller thrust to the ship. These are the *gearbox*, if fitted, the propeller shafting, the various bearings and the *propeller*.

transmitter A device which receives a measurement signal and produces a related output signal. This is then transmitted some distance to a *controller*.

transom The aftermost athwartships region of a vessel which gives a flat stern shape.

211

transom stern   A flat plate stern construction of almost triangular section. It is simpler to construct when compared with a *cruiser stern* and also gives an increased after deck area. See Figure T.6.

transverse   (1) The direction at right-angles to the centreline of a ship. (2) An item of structure at right-angles to the centreline of a ship. See Figure F.4.

transverse metacentre   See *metacentre*.

transverse thrust   A thrust which is at right-angles to the motion of the ship. It may refer to the side thrust created by a single propeller when moving a ship forward.This is particularly evident when manoeuvring close to a harbour wall.

trapezoidal rule   A means of determining the area under a curve by approximating it to a straight line. A number of equally spaced ordinates are then used to form trapezoids whose area can be found.

travelling block   An assembly of sheaves through which wires from the *draw-works* are passed. A hook at the bottom is attached to the *swivel* and hence the *drill string,* which can be raised or lowered.

tree   See *christmas tree*.

triac   A *semiconductor* device made up of two *silicon controlled rectifiers* connected back-to-back on a single piece of silicon.

trial condition   The loaded condition specified for a vessel when on *sea trials.*

trial trip   A short voyage, for a recently completed ship, in which the machinery and equipment is tested to ensure compliance with the contract specifications. Owner's and builder's representatives and classification society surveyors will be present.

trial trip rating   A short time *rating* for an engine when on sea trials.

tribology   The study of interacting surfaces in relative motion. It includes friction, lubrication and wear and relates to bearings, gears, etc.

trickle charging   A means of maintaining a *battery,* which is not in use, in a fully charged condition. A small current is supplied which is equal to the internal losses of the battery.

trim   The inclination of a vessel in a fore and aft direction. When the draughts forward and aft are the same, the vessel is said to be on an even keel. If the draught aft is greater, then the vessel is trimmed by the stern. If the draught forward is greater, then the vessel is trimmed by the head.

triple point of water   The temperature at which ice, water and water vapour are all in equilibrium at a pressure of one standard atmosphere in a sealed vacuum flask (triple point cell). See *kelvin scale.*

tripping   (1) The action of drilling a hole. See *round trip.* (2) The opening of an electric switch. See *preferential tripping.*

tripping bracket   A flat bar or plate fitted to a deck girder, stiffener, beam, etc., to reinforce the free edge.

trunk   A passage extending through one or more decks to provide access or ventilation to a space.

trunk line   The main feeder line for oil or gas.

trunk piston engine An internal combustion engine in which the connecting rod is directly connected to the piston by a gudgeon pin. See Figure T.7.

tube plate A thick metal plate into which tubes are secured as in a *shell and tube heat exchanger*.

tubes The heat exchange surfaces in *condensers, shell and tube heat exchangers, boilers,* etc. Lengths of appropriate material are secured between tube plates in heat exchangers, drums or headers in boilers, etc.

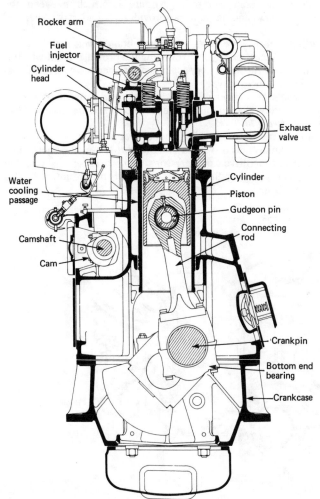

Figure T.7 A trunk piston engine

213

tufnol   A laminated plastic material with good insulating properties. It is often used for bearings, in particular, *stern tubes* and *rudder pintles*.

tumblehome   An inward curvature of the midship side shell, in the region of the upper deck. See Figure P.7.

tungsten   A hard, grey, metal which is resistant to corrosion and used in high speed steel alloys for drill bits and cutting tools.

tungsten inert gas welding   A welding process for thin sheet using a non-consumable, tungsten electrode. An arc is struck between the electrode and the plate and an inert gas provides a shield and a filler rod the weld metal.

tuning   The adjustment of an electric circuit to obtain *resonance*.

tunnel   A watertight access passage surrounding the propeller shaft which is fitted on a ship where the machinery space is positioned towards midships.

tunnel escape   See *means of escape*.

turbine   A machine which produces rotary motion as a result of a fluid striking appropriately shaped blades on a wheel. See *impulse turbine, reaction turbine, steam turbine*.

turbo feed pump   A *steam turbine* driven boiler *feed pump*.

turbocharger   An exhaust gas driven air compressor. It uses the energy in a diesel engine exhaust to drive a compressor which provides pressurized air for *scavenging* and charging the engine. The term turbo-blower is also used. See Figure T.8.

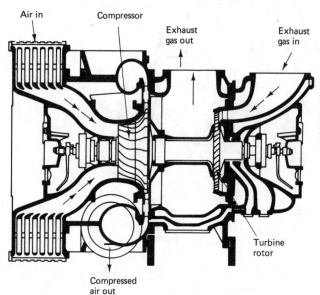

Figure T.8 A turbocharger

214

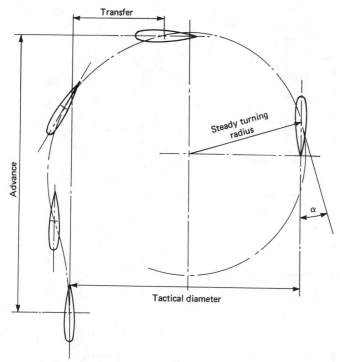

Figure T.9 A turning circle

turbulence stimulator A device fitted to a ship model being tested which will ensure that laminar flow does not occur on the hull surface. A trip wire or small projecting studs may be fitted at the forward end. See *tank testing*.

turbulent flow Fluid flow where the particle motion at any point is rapidly changing both in direction and magnitude. It occurs at high *Reynolds numbers*.

turn-down ratio The ratio of maximum to minimum flow that can take place through a device, e.g. a boiler burner.

turning circle A circle moved through by a ship when the rudder is placed in its extreme position. It is a manoeuvre carried out on *sea trials*. See Figure T.9.

turning gear A reversible electric motor which, through a system of gears, can be used to slowly turn a large diesel engine or steam turbine and gearbox assembly. It enables precise positioning for overhaul or examination.

215

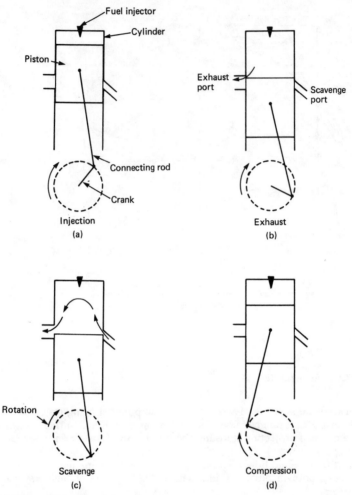

Figure T.10 The two stroke cycle: (a) injection, (b) exhaust, (c) scavenging, (d) compression

tween decks   The upper cargo stowage compartments, or the space between any two adjacent decks.

twin screw vessel   A vessel filled with two propellers, one on either side of the centreline. The starboard propeller will turn clockwise, when viewed from aft, and is right-handed, the port propeller anti-clockwise and is left-handed.

two-step controller A *controller* whose output signal changes from one predetermined value to another as a result of the deviation changing sign. See *on-off action*.

two-stroke cycle An operating cycle for an internal combustion engine which requires two strokes or one revolution of the crankshaft. See Figure T.10.

two-term controller A *controller* which provides *proportional action* and either *integral* or *derivative action*.

U

ullage The distance between the ullage lip at the top of a tank and the liquid surface below.

ultimate tensile stress The highest stress value occurring in a material during a *tensile test*. It is obtained by dividing the load by the original cross-sectional area of the test piece.

Ultra Large Crude Carrier An *oil tanker* with a deadweight in the region of 500 000 tonnes. See *Very Large Crude Carrier*.

ultra long stroke engine A slow-speed, two-stroke engine which uses a large stroke to bore ratio to utilize the improved thermodynamic efficiency resulting from uniflow scavenging. It also uses a lower shaft speed to improve propeller efficiency.

ultra-violet light crack detection A *dye penetrant testing* method to detect surface cracks. A fluorescent penetrant is used and can be detected using an ultra-violet light. See *non-destructive testing*.

ultra-violet light fluorescence The emission of light from a molecule that has absorbed light. During the interval between absorption and emission, energy is lost and light of a longer wavelength is emitted. Oil fluoresces more than water and this becomes a means of detecting it.

ultrasonic testing The use of high frequency sound waves to test a material. They are reflected back from the far side and can be displayed on a cathode ray oscilloscope. A defect will be shown and its size and location may be found. See *non-destructive testing*.

umbilical A flexible cable or pipe which connects equipment or a diver to a control position in a ship or rig and supplies power, life support or other services.

unattended machinery space A classification society *notation* indicating that certain essential operational and safety requirements have been met. The ship may operate for certain periods with the machinery space unattended.

unbalanced rudder A *rudder* design in which no part of the area is forward of the turning axis. See Figure U.1.

undamped natural frequency The *natural frequency* of oscillation of a system that would occur if damping were reduced to zero.

undercooling The reduction of condensate temperature below the steam temperature in a *condenser*.

underdamping A degree of *damping* in a system which is so small that when a disturbance occurs, one or more cycles of oscillation occur.

undervoltage release A device which will operate to open a *circuit breaker* if the voltage falls below a particular pre-set value. It also prevents a circuit breaker being closed if the generator voltage is low or non-existent.

underway A ship which is not at anchor or in any way secured to the shore

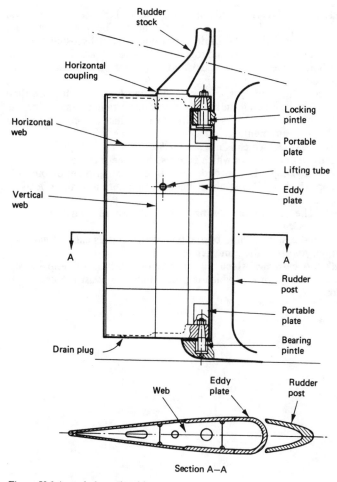

Section A–A

Figure U.1 An unbalanced rudder

or aground. In a general sense it is considered to be a ship moving through the water.

underwriter A person who insures the whole or a part of a ship or its cargo. See *Lloyd's Corporation*.

uniflow scavenging A *scavenging* system where the incoming air enters at the lower end of the cylinder and leaves at the top. The outlet at the top of the cylinder may be ports or a large valve.

union purchase rig A cargo handling arrangement which uses two derricks, one over the quayside and one over the hold. The wires from both derricks are shackled to the same cargo hook. Thus by using the two winch controllers, separately and together, the hook is raised and lowered over the hold, travels over the deck and can be raised and lowered over the ship's side.

unit (1) A complex built-up section of a ship which can weigh more than 100 tonnes, its size being limited by the transportation capacity of the shipyard's equipment. (2) Any quantity or value which is used as a basis or standard of measurement.

unseaworthy A vessel which is not *seaworthy*.

unstable Any item or system which is not stable. See *stability*.

unsymmetrical flooding The entry of water into a compartment on one side of the ship which will bring about heeling, in addition to the effects of flooding.

upper deck The uppermost continuous deck where more than one continuous deck exists.

uptake A metal casing or large bore piping which carries exhaust gases up through the *funnel* to the atmosphere.

user friendly A *computer* which will operate on the basis of simple easily understood instructions or which is programmed to assist the user in its operation.

V

Vac-Strip cargo pumping system The use of a separator tank, vacuum pump and associated equipment in conjunction with a centrifugal pump to enable high capacity pumping and stripping of cargo oil. Non-liquid elements are separated from the oil before it reaches the pump suction and the pump discharge rate is varied to match the oil flow to the pump suction.

vacuum distillation An oil refining process in which heavy distillates are boiled under reduced pressure and subsequently lower temperatures.

valve An assembly which is fitted into a pipeline to control the flow of fluid. See *butterfly valve, globe valve, non-return valve, quick closing valve, relief valve.*

valve bridge The part of a *valve* assembly through which the spindle is screwed to raise or lower the valve disc. It may also be called the *yoke.*

valve cage A cylinder fitted with ports in which a *valve plug* moves. The port openings are shaped to produce various flow characteristics for different valves, e.g. linear or equal-percentage.

valve characteristic The relationship between valve lift and flow. It is of particular importance with *control valves.* The common characteristics are quick-opening, linear and equal percentage (semi-logarithmic). See Figure V.1.

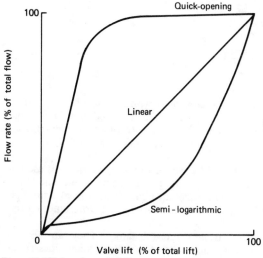

Figure V.1 Valve characteristics

221

valve chest A series of valves built into a single block or manifold. Various arrangements of suction and discharge connections are possible with this assembly.

valve disc A movable cover which provides a variable restriction to fluid flow. A *control valve* may have two discs or *valve plugs*, which are specially shaped to create the *valve characteristic*.

valve land The part of the spool of a *spool valve* which cuts off the flow of fluid through a port by covering it.

valve lid See *valve disc*.

valve plug See *valve disc*.

valve rotator A device which slowly rotates inlet and exhaust valves on four-stroke engines. *Rotocap* is a particular design.

vane pump An assembly in which a rotor runs eccentrically within a casing. Several vanes are located in slots within the rotor and are free to slide radially. As the rotor turns vanes move against the casing and pump liquid. See Figure V.2.

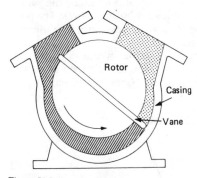

Figure V.2 A vane pump

vanning The loading or stuffing of *containers* with goods to be shipped.

vaporizing liquid A fire extinguishing agent which is used in both portable and fixed fire fighting equipment. Halon 1301 (bromotrifluoromethane, BTM) and Halon 1211 (bromochlorodifluoromethane, BCF) are two commonly used liquids.

vapour compression cycle A form of reversed *Carnot cycle* which is used in refrigeration systems. The refrigerent gas is compressed and then condensed in a condenser. The refrigerant liquid is then passed through an expansion valve and evaporates in an evaporator before entering the compressor.

vapour lines The vent pipes from cargo oil tanks. They are led to *pressure/vacuum valves* which are usually mounted on standpipes some distance above the deck.

vapour pressure The pressure exerted by a vapour either alone or in a mixture of gases.

vapour pressure thermometer A *temperature* measuring instrument which uses a sensing bulb which is partly filled with a volatile liquid such as methylchloride or diethylether. The remainder of the bulb, the capillary tubing and the *Bourdon tube* are filled with the liquid vapour. The measuring range for methylchloride is 0 to 50°C and diethylether 60 to 100°C.

variable area flowmeter An instrument which measures the volume flow rate by reference to the area of a constriction in a pipeline, when the pressure differential is held constant. See *rotameter*.

variable stroke pump A *Hele-Shaw* or *swashplate pump* in which the crank throw or swash angle respectively can be varied so that the amount of fluid delivered per revolution of the pump can be varied.

velocity compounding In an *impulse turbine,* this is the use of a single nozzle with an arrangement of several moving blades on a single disc. Between the moving blades are fitted guide blades which are fastened to the casing. The steam velocity is thus progressively reduced through the turbine.

vena contracta The position at which liquid flowing through an orifice contracts to a minimum area downstream. It is the minimum static pressure position and also the maximum velocity point.

ventilation The circulation and refreshing of the air in a space without necessarily a change of temperature.

ventilator Any grid, vent or opening through which air can enter or leave a space. A means of permanent closure is required by the Load Line Rules for all except high ventilators.

ventilator head A cover arrangement which is fitted over a *ventilator* to prevent the entry of rain, or sea water. It may also be used as a means of closure. See *cowl*.

venturi A convergent–divergent duct in a tube or pipe which creates a change in pressure and velocity of the flowing liquid and enables a measurement of flow rate.

vertical keel See *centre girder*.

vertical lift control The use of foils to raise a vessel out of the water, as in hydrofoil craft.

Very Large Crude Carrier An *oil tanker* with a deadweight in the region of 200 000 to 500 000 tonnes. See *Ultra Large Crude Carrier*.

vessel Any craft bigger than a rowing boat and used in navigation. It generally includes any water-borne craft, other than a seaplane, which is, or can be, used as a means of transportation on water.

vibrating reed tachometer A *tachometer* which uses a set of thin reeds, each with a different natural frequency, and fixed at one end. When placed in contact with a rotating machine one reed will vibrate and the speed is read from a scale.

vibration A type of oscillatory motion of a body when displaced from an equilibrium position by an external force. The maximum displacement is called the amplitude and the time interval between cycles is called the period. See flexural vibration, torsional vibration.

vibration damper A device fitted to an engine crankshaft to suppress or reduce the stresses resulting from *torsional vibration*. See *detuner*.

vibration measurement The determination of *frequency* and amplitude of *vibration*, usually as a means of diagnosing a problem with rotating machinery before damage occurs.

vibration post A projection on a *stern frame*, above the arch, which provides a connecting point to a floor in the after end structure of the ship.

Victory ship A particular design of *Liberty ship*.

virtual mass factor A value used, in conjunction with the *displacement* of a ship, to determine the mass of water and vessel which is set in motion when *heaving, pitching, rolling* or *vibrating*. See *added virtual mass*.

vis-breaking Viscosity breaking. A means of reducing the *viscosity* of the light end, i.e. fuel oil, of a heavy *fraction*.

viscometer An instrument which measures the *viscosity* of an oil. *Redwood, Saybolt* and Engler instruments are used and give flow times for a particular volume of fuel. A capillary tube device is used for monitoring of fuel oil supplies to boilers and diesel engines and is usually part of a control system to maintain a desired value. See Figure V.3.

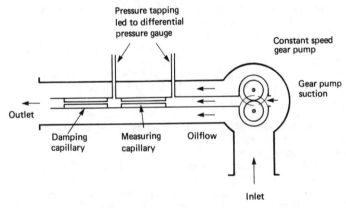

Figure V.3 A viscometer

viscosity The resistance to flow offered by a liquid. A free flowing liquid will have a low viscosity whilst a heavy oil will be viscous and have a high viscosity.

visual (video) display unit An output device for a *computer*, usually a television screen.

Voith Schneider propeller A propulsion device which can provide a variable directional thrust to enable steering. It consists of a vertical rotating disc

224

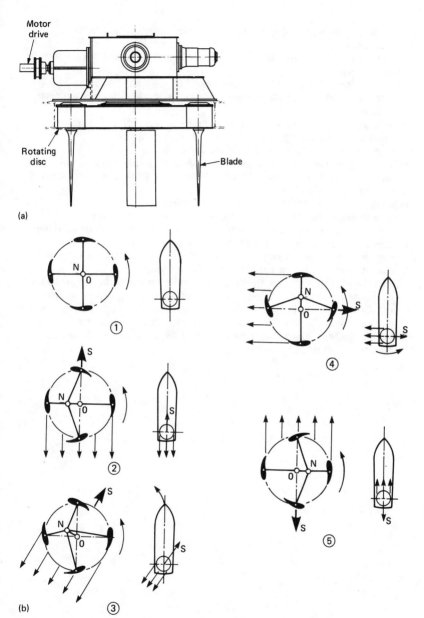

(a)

Motor drive

Rotating disc

Blade

(b)

Figure V.4 A Voith Schneider propeller: (a) construction, (b) operation

225

containing vertical blades around its periphery. The blades can be independently rotated to produce the directional thrust. See Figure V.4.

volatile memory   A computer *memory* whose contents are lost when power is disconnected.

volt   The derived unit for the measurement of electrical potential.

voltage   The value of an *electromotive force* or *potential difference* expressed in *volts*.

voltage dip   An instantaneous reduction in output voltage from an a.c. *generator* when a large load current is suddenly supplied.

voltage regulator   See *automatic voltage regulator*.

voltage transformer   An *instrument transformer* which reduces the high mains voltage to a low value for a *voltmeter* or the voltage operated coils of instruments and *relays*.

voltmeter   An instrument which measures *voltage*.

volume of displacement   The volume of water displaced by a ship.

volute   A spiral shape. It is the internal casing shape of a *centrifugal pump*.

voyage   A journey by sea to a distant place or country.

voyage charter party   A contract for the hire of a vessel for the carriage of specific goods between certain ports, i.e. a voyage.

Vulcan clutch (coupling)   A hydraulic unit which connects the engine to the propeller shafting. A driving section or impeller and a driven section or runner are housed in a casing. Rotation of the impeller causes a circulation of oil between the two and transmission of the drive.

vulcanite   *Rubber* which has been hardened by treatment with sulphur at high temperature, i.e. vulcanized.

W

waist   The midship section of a ship between the *forecastle* and the *poop*.

waiting on weather   A delay due to bad weather.

wake   The water which is in motion at the stern of a ship as a result of a ship's movement.

wake fraction   The ratio of the wake speed to the *speed of advance* or the ship speed. The Froude wake fraction uses the speed of advance and the Taylor wake fraction the ship speed.

wall sided formula   A formula relating the *righting lever*, the *metacentric height* and the angle of inclination for a vessel of rectangular cross-section, when inclined up to about ten degrees.

Ward-Leonard speed control   A means of speed or directional control for a d.c. motor which is supplied by a d.c. generator. The d.c. generator is driven by an a.c. or d.c. motor and its armature voltage is varied or reversed by control of the field current. See Figure W.1.

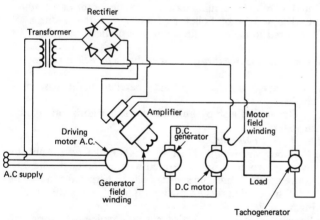

Figure W.1 Ward Leonard speed control

warp end   A shaped drum fitted on the end of a *winch* drum axle. It is used to move a ship using ropes or wires which are fastened to *bollards* ashore. See Figure W.4.

wash   The waves and ripples spreading out from a vessel when under way.

wash bulkhead   A perforated *bulkhead* fitted into a cargo or deep tank to reduce the sloshing or movement of liquid through the tank.

waste heat   The energy from combustion in an internal combustion engine which is not converted into useful work, e.g. energy in the exhaust gas or cooling water.

waste heat boiler A *boiler* which uses exhaust gas from an engine to produce low pressure saturated steam.

watch A time period, usually of four hours, e.g. 12–4, 4–8, 8–12, which operates around the clock. It is the working period for one or more officers and crew in the navigation and engineering departments.

water drive The use of a natural or pumped flow of water into a *reservoir* to maintain the pressure and drive out the oil or gas during production.

water gauge A direct reading water level measuring device, usually a glass tube, which is fitted to a vessel containing water. See *gauge glass*.

water hammer The impact of moving water in pipework which is stopped suddenly. It may occur when steam is admitted to a pipeline containing a reasonable surface of water. The steam condenses, a partial vacuum occurs and draws the water along the pipe until it strikes a bend or valve.

water pocket A place where water collects as a result of poor design.

water pressure test (1) The testing of a *tank* or *bulkhead* by a maximum service pressure head of water. (2) The testing of a pressure vessel by filling with water and pressurizing to some maximum value.

water ring primer An air pump which is used to prime *centrifugal pumps*. It may be mounted on top of the pump or as a separate motor driven unit.

water treatment The testing of boiler *feed water* and the addition of chemicals as required to maintain the water in a particular condition of purity.

waterjet propulsion The use of a *thruster*, rather than a *propeller*, to provide propulsion.

waterline An imaginary line on a ship's side parallel to the base at a particular draught.

waterplane area The area of a ship's hull at a particular horizontal plane, i.e. within the waterline.

watertight bulkhead See *bulkhead*.

watertight door A door which is fitted in a watertight bulkhead and able to open vertically or horizontally. It is operated by a hydraulic mechanism either locally or remotely. It must be substantially constructed and able to withstand the full hydraulic pressure of the adjoining compartment if it floods. See Figure W.2.

watertight subdivision The dividing up of a ship's hull into a series of watertight compartments by means of watertight *bulkheads*. See *floodable length*.

watertube boiler A high pressure, high temperature steam supplying *boiler*. The *feed water* is heated within tubes by hot gases passing over them.

waterwall tube A tube used in the wall of a *boiler* furnace which acts as a heat transfer surface and also part of the casing, thus minimizing the need for insulation. See *membrane tube, tangent tube*.

waterwashing The spraying of fresh water onto the gas turbine blades of a *turbocharger* to clean off combustion deposits.

228

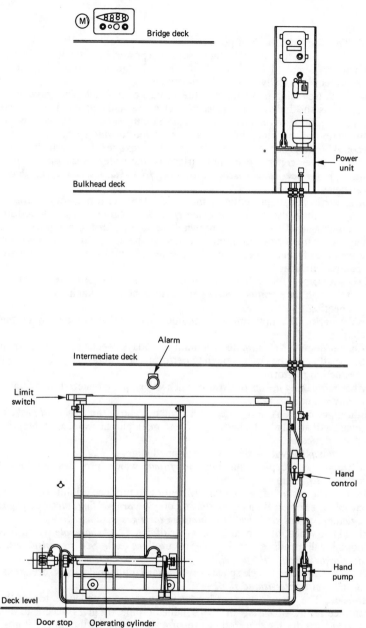

Bridge deck

Bulkhead deck

Power unit

Intermediate deck

Alarm

Limit switch

Hand control

Hand pump

Door stop   Operating cylinder

Deck level

Figure W.2 A watertight door

watt The derived *SI unit* of power.

wattmeter An instrument which measures the power, in watts, in an electric circuit or supplied by an a.c. generator. If the scale is graduated in kilowatts it is usually called a kilowattmeter.

wave The travelling undulation of the surface of a liquid. The wave crest moves as the individual surface particles rise and fall but there is no forward motion of the liquid. Waves arise on the surface of the sea due to the action of the wind and their size is related to wind speed.

wave-making resistance The component of a ship's resistance which is due to the waves generated on the surface of the water as the ship moves through it. See *eddy-making resistance, frictional resistance, residuary resistance.*

wave spectra A data presentation method for the energy present in *waves* in a sea. It may also be called a *sea spectra*. A curve of wave spectrum ordinate against frequency is drawn. The area under the curve between any two frequencies represents the sum of the energy of all component waves with frequencies in the range. Several formulae exist for the wave spectrum ordinate.

weardown gauge A depth gauge which is inserted into the *stern frame* boss and through the *sterntube bearing* to measure the wear that has occurred in the bearing.

weather deck The uppermost continuous *deck* which is exposed to the elements.

weather window An estimate of a number of days, weeks or months when the weather will be good enough to perform a critical task, e.g. installing a production platform.

weathertight Any means of closure which will stop the penetration of water in any sea condition.

weathertight door A door fitted in a structure above the *freeboard deck*. It must be of adequate strength and able to maintain the watertight integrity of the structure.

web A flat plate with a flanged or stiffened edge.

web frame A deep-section built-up *frame* which provides additional strength to a ship's structure.

welding The fusion of two metals by heating to produce a joint which is as strong or stronger than the parent metal. See *arc welding, metal inert gas welding, oxy-acetylene welding, tungsten inert gas welding.*

well (1) A space into which bilge water drains. (2) A *borehole* which it is hoped will produce, or is able to produce, oil. See *appraisal well, development well, wildcat well.*

well deck The open deck space between raised upper decks or erections, usually on a *three-island ship*.

well-head The various items of hardware which are installed on top of the surface string of *casing* or the conductor pipe. It gives access to the *borehole* for drilling, controlling pressure and regulating the flow of fluids.

wet bulb temperature The *temperature* measured by a thermometer which has its bulb kept moist by a water soaked wick. See *dry bulb temperature, psychrometer*.

wet liner A *cylinder liner* which has cooling water in contact with its outside surface. It is sealed against leakage at either end where it fits into the *cylinder block*.

wet steam Steam which contains some fine particles of water. See *saturated steam, steam*.

wet sump The *crankcase* of an internal combustion engine which is used as a tank to store lubricating oil.

wetted surface area The area of a ship's hull which is in contact with the water.

wharf See *quay*.

Wheatstone bridge An instrument which measures electrical *resistance*. It consists of four resistors, one of which is unknown, connected as a *bridge* or box. A supply is provided to opposite corners and a *galvanometer* is connected across the other two. The bridge is balanced when the galvanometer reads zero and the unknown resistor can be calculated. See Figure W.3.

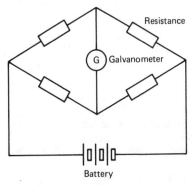

Figure W.3 A Wheatstone bridge

wheel The steering wheel which controls the position of the *rudder*.

wheelhouse See *bridge*.

whipstock A wedge-shaped tool which deflects the angle of the drill *bit* in order to bring about *deviated drilling*.

whirling The transverse *vibration* of a rotating shaft in *resonance*.

whistle A device which is designed to produce audible signals in the form of short and prolonged blasts. It may be operated by steam or compressed air. It is required by international regulations.

whistle stop valve A small bore *non-return valve* which supplies the whistle with steam direct from the boiler.

white metal A tin based alloy with amounts of lead, copper and antimony. It may also be a lead based alloy with antimony. It has a low coefficient of friction and is used as a lining material for bearings.

wildcat well An *exploration* well drilled in search of a new oil or gas *reservoir*.

winch A machine which utilizes the winding or unwinding of rope or wire around a barrel for various cargo handling or mooring duties. See Figure W.4.

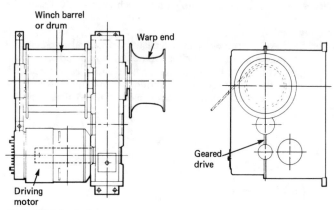

Figure W.4 A winch

windage loss The power loss in a *steam turbine* due to the astern turbine blading churning steam within the casing.

windlass A machine used for hoisting and lowering the anchor. See Figure W.5.

windward The direction from which the wind is blowing.

Winter North Atlantic A *load line* marking which is permitted for a ship of 100 m length or less operating in the North Atlantic. See Figure L.5.

wire drawing An effect produced by the throttling of a fluid, e.g. steam, through a small orifice such as an almost closed valve.

wireline A wire or cable used in a *borehole*.

wiring diagram A drawing which shows the detailed wiring and connections between items of electrical equipment and may also show the routing of connections.

woodruff key A semi-circular *key*.

word The unit of information which can be transmitted, stored and operated upon at one time in a *computer*.

word processing The manipulation of written material, e.g. letters, reports, memos, etc., by a *computer* program.

232

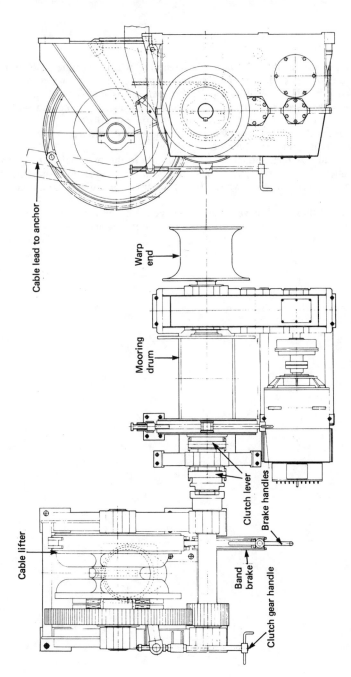

Figure W.5 A windlass

Cable lead to anchor

Warp end

Mooring drum

Cable lifter

Clutch lever

Brake handles

Band brake

Clutch gear handle

work measurement A *work study* technique which examines the human effort or work content of a job with a view to improving *productivity*.

work study The use of various techniques, in particular work study and *work measurement,* to critically examine all aspects of human work in order to improve the efficiency and economy of effort.

workover A programme of work involving re-entry into a completed *well* to do modification or repair work.

worm gear A high reduction ratio gear where the two shafts are at right angles. A cylindrical core, or worm, with a single or multi-start helical gear, drives a toothed wheel. It is often used for a diesel engine *turning gear.*

wreck A ship which is very badly damaged and unseaworthy.

write To transfer information into a computer *memory.*

X

X-rays Electromagnetic radiation in the form of rays of short wavelength. They are able to penetrate matter which is opaque to light. See *radiography*.

Y

Y-connection See *star connection*.

yawing The motion of a ship when it is rotating about a vertical axis. See Figure R.6.

yield point See *elastic limit*.

yield stress The value of stress at which a metal under *tensile test* undergoes plastic deformation. A considerable amount of deformation occurs with iron and annealed steels.

yoke (1) The ferromagnetic material in a magnetic circuit which is not surrounded by electrical windings. (2) See *valve bridge*.

York Antwerp Rules 1974 A set of rules for the adjustment of *general average*. They are accepted by most maritime nations.

Young's modulus See *modulus of elasticity*.

Z

zener diode A *diode* across which the voltage drop is constant over a range of current. It is used as a means of providing a stabilized voltage in some circuits.

zero The *datum* point for many measuring scales and the required reading for *null measurement*.

zero error The reading on an *instrument* when it is disconnected from the measured parameter and does not indicate zero.

zinc A hard, white metallic element with a good resistance to atmospheric corrosion. It is used as a coating for steel in a process called *galvanizing* and also as an alloying element.

Appendix

SI units

The metric system of units, which is intended to provide international unification of physical measurements and quantities, is referred to as SI (Système International).

There are three classes of units: base, supplementary and derived. There are seven base units: length in metres (m); mass in kilograms (kg); time in seconds (s); electric current in amperes (A); temperature in kelvins (K); luminous intensity in candelas (cd); and amount of substance in moles (mol). There are two supplementary units: plane angle in radians (rad); solid angle in steradians (sr). All remaining units used are derived from the base units. The derived units are coherent in that the multiplication or division of base units produces the derived unit. Examples of derived units are given in Table 1.

TABLE 1. Derived units

Quantity	Unit
Force	Newton (N) = $kg\,m/s^2$
Pressure	Pascal (Pa) = N/m^2
Energy, work	Joule (J) = $N\,m$
Power	Watt (W) = J/s
Frequency	Hertz (Hz) = $1/s$

There are in use certain units which are not SI but are retained because of their practical importance. Examples are: time in days, hours, minutes, and speed in knots.

To express large quantities or values a system of prefixes is used. The use of a prefix implies a quantity multiplied by some index of 10. Some of the more common prefixes are:

1 000 000 000	=	10^9	=	giga	=	G
1 000 000	=	10^6	=	mega	=	M
1 000	=	10^3	=	kilo	=	k
100	=	10^2	=	hecto	=	h
10	=	10^1	=	deca	=	da
0.1	=	10^{-1}	=	deci	=	d
0.01	=	10^{-2}	=	centi	=	c
0.001	=	10^{-3}	=	milli	=	m
0.000 001	=	10^{-6}	=	micro	=	μ
0.000 000 001	=	10^{-9}	=	nano	=	n

Example:

10 000 metres	=	10 kilometres	=	10 km
0.001 metre	=	1 millimetre	=	1 mm

Note: Because kilogram is a base unit care must be taken in the use and meaning of prefixes. Only one prefix can be used, for example, 0.000 001 kg = 1 milligram.

A conversion table for some well known units is provided in Table 2.

TABLE 2 Conversion factors

To convert from	To	Multiply by
Length		
inch (in)	metre (m)	0.0254
foot (ft)	metre (m)	0.3048
mile	kilometre (km)	1.609
nautical mile	kilometre (km)	1.852
Volume		
cubic foot (ft^3)	cubic metre (m^3)	0.02832
gallon (gal)	litre (l)	4.546
Mass		
pound (lb)	kilogram (kg)	0.4536
ton	kilogram (kg)	1016
Force		
pound-force (lbf)	newton (N)	4.448
ton-force	kilonewton (kN)	9.964

TABLE 2 Conversion factors (cont.)

To convert from	To	Multiply by
Pressure		
pound-force per square inch (lbf/in^2)	kilonewton per square metre (kN/m^2)	6.895
atmosphere (atm)	kilonewton per square metre (kN/m^2)	101.3
kilogram force per square centimetre) (kgf/cm^2)	kilonewton per square metre (kN/m^2)	98.1
Energy		
foot pound-force	joule (j)	1.356
British thermal unit (Btu)	kilojoule (kJ)	1.055
Power		
horsepower (hp)	kilowatt (kW)	0.7457
metric horsepower	kilowatt (kW)	0.7355

Abbreviations

A	ampere
Å	ångstrom
A1	Lloyd's classification, first class
a.a.	always afloat
a.a.r.	against all risks
AB	able bodied seaman
ABS	American Bureau of Shipping
abs.	absolute
a.c.	alternating current
ACV	air cushion vehicle
ad val.	ad valorem
a.f.	advance freight
Ah	ampere hour
ALU	arithmetic and logic unit
amp.	amplitude
API	American Petroleum Institute
atm.	atmospheric
a.v.r.	automatic voltage regulator
b	breadth, barrel
BASIC	Beginner's All-purpose Symbolic Instruction Code
bbl	barrel

BCF	bromochlorodifluoromethane
bd	barrels per day
b.d.c.	bottom dead centre
BICERA	British Internal Combustion Engine Research Institute
BMEC	British Marine Equipment Council
BOP	blowout preventer
b.p.	brake power
Brl	barrel
BSI	British Standards Institution
BSRA	British Ship Research Association
BTM	bromotrifluoromethane
Btu	British thermal unit
BV	Bureau Veritas
C	capacitance
°C	degree Celsius (centigrade)
C & F	cost and freight
C & I	cost and insurance
CALM	catenary anchor leg mooring
CBM	conventional buoy mooring
CSC	computer supervisory control
CEng	Chartered Engineer
CFS	container freight station
c.g.	centre of gravity
CIF	cost, insurance and freight
c.m.r.	continuous maximum rated
COBOL	COmmon Business Oriented Language
coeff	coefficient
const.	constant
corr.	corrected
COW	crude oil washing
CPA	critical path analysis
CPM	critical path method
c.p.p.	controllable pitch propeller
CPU	central processing unit
crit.	critical
c.r.t.	cathode ray tube
CSC	computer supervisory control
c.s.r.	continuous service rating
CTL	constructive total loss
d	diameter
d and a	dry and abandoned
dB	decibel
DB	double bottom
DBS	Distressed British Seaman

d.c.	direct current
DDC	direct digital control
Dep	departure
DIN	Deutsche Industrie-Normen
DNV	Det Norske Veritas
DS	drill ship
DT	deep tank
dwt	deadweight
ELSBM	exposed location single buoy mooring
e.m.f.	electromotive force
e.p.	effective power
EPROM	erasable programmable read-only memory
ER	engine room
ETA	estimated time of arrival
F	farad, fresh
°F	degree Fahrenheit
FCL	full container load
FD	free delivery, forced draught
f.i.o.	free in and out
f.m.	frequency modulation
FO	fuel oil
FOB	free on board
FOC	free of charge
FORTRAN	Formula Translation
FPT	fore peak tank
ft	feet
FW	fresh water
g	gram
gal	gallon
GL	Germanischer Lloyd
GMT	Greenwich Mean Time
gr	gross
g.r.p.	glass reinforced plastic
GRT	gross register tonnage
H	magnetizing force, magnetic field strength
h	height
HCV	higher calorific value
HMS	Her Majesty's Ship
hp	horse power
HP	high pressure
Hz	hertz (frequency)

I	current
IACS	International Association of Classification Societies
IC	integrated circuit
ICS	Institute of Chartered Shipbrokers
IGS	inert gas system
IMCO	now IMO
IMO	International Maritime Organization
in	inch
INMARSAT	International Maritime Satellite Organization
i/o	input/output
i.p.	indicated power
IP	Institute of Petroleum
i.r.	insulation resistance
IS	intrinsically safe
ISO	International Organization for Standardization
JU	jack-up
k	kilo ($\times 10$)
K	kelvin
kW	kilowatt
L	inductance
l	litre, length
LASH	lighter aboard ship
lb	pound
LCD	liquid crystal display
l.c.l.	less than container load
LCV	lower calorific value
LED	light emitting diode
LEL	lower explosive limit
LMC	Lloyd's Machinery Certificate
LNG	liquefied natural gas
LO	lubricating oil
LoLo	lift-on lift-off
LPG	liquefied petroleum gas
LR	Lloyd's Register
LRMC	Lloyd's Refrigerating Machinery Certificate
lub.	lubricating
MARPOL	Marine Pollution Convention
max.	maximum
m.c.b.	miniature circuit breaker
MEP	mean effective pressure
m.i.c.c.	mineral insulated copper cable
MIG	metal-inert-gas

min.	minimum
MM	mercantile marine
m.m.f.	magnetomotive force
MN	merchant navy
MOT	Ministry of Transport
MS	motor ship
MSA	Merchant Shipping Act
MSV	multi-function support vessel
MUF	make-up feed
mV	millivolt
MV	motor vessel
NDT	non-destructive testing
NPL	National Physical Laboratory
NRT	net register tonnage
n.t.p.	normal temperature and pressure
O & M	operations and methods
o.a.	overall
OBO	oil bulk ore
P & I	protection and indemnity
PERT	programme evaluation and review technique
p.f.	power factor
pH	hydrogen ion exponent
PROM	programmable read-only memory
ptfe	polytetrafluorethylene
PV	pressure-volume
PVC	polyvinyl chloride
rad/s	radian per second
RAM	random-access memory
rev/min	revolutions per minute
RN	Royal Navy
ROM	read-only memory
RoRo	roll-on roll-off
r.p.m.	revolutions per minute
S	summer
SALM	single anchor leg mooring
SBM	single point buoy mooring
SBT	segregated ballast tanks
SCR	silicon controlled rectifier
SDNR	screw down non-return
SFC	specific fuel consumption
SI	International System of Units

SL	screw lift
SOLAS	safety of life at sea
s.p.	shaft power
SPBM	single point buoy mooring
sp.gr.	specific gravity
sp.ht.	specific heat
SPM	single point mooring
sp.vol.	specific volume
SR & CC	strikes, riots and civil commotions
SS	steam ship
STCW	Standards of Training Certification and Watchkeeping for Seafarers
s.t.p.	standard temperature and pressure
str	steamer
SVP	saturation vapour pressure
SW	sea water
SWATH	small waterplane area twin hull
SWL	safe working load
SWOPS	single well oil production system
T	torque, tropical
t	time
TBN	total base number
TBO	time between overhauls
t.d.c.	top dead centre
temp.	temperature
TF	tropical fresh water load line
T/L	total loss
TPC	tonnes per centimetre
TR	tons registered
UD	upper deck
UEL	upper explosive limit
ULCC	ultra large crude carrier
UMS	unattended machinery space
UV	ultra-violet
V	volt
VA	volt-ampere
vac.	vacuum
VAr	volt-amperes reactive
v.d.	vapour density
VDU	visual (video) display unit
VI	viscosity index
VLCC	very large crude carrier
vol.	volume

v.p.	vapour pressure
W	watt, winter
Wh	watt hour
WNA	Winter North Atlantic
WOW	waiting on weather
wt	weight
Z	impedance